Basic Probability Theory
for Biomedical Engineers

© Springer Nature Switzerland AG 2022

Reprint of original edition © Morgan & Claypool 2006

Basic Probability Theory for Biomedical Engineers
John D. Enderle, David C. Farden, Daniel J. Krause

ISBN: 978-3-031-00485-8   paper

ISBN: 978-3-031-01613-4   ebook

DOI: 10.1007/978-3-031-01613-4

Library of Congress Cataloging-in-Publication Data

A Publication in the Springer series
SYNTHESIS LECTURES ON BIOMEDICAL ENGINEERING
Lecture #5
Series Editor and Affliation: John D. Enderle, University of Connecticut

1930-0328 Print
1930-0336 Electronic

First Edition
10 9 8 7 6 5 4 3 2 1

Basic Probability Theory for Biomedical Engineers

John D. Enderle
Program Director & Professor for Biomedical Engineering
University of Connecticut

David C. Farden
Professor of Electrical and Computer Engineering
North Dakota State University

Daniel J. Krause
Emeritus Professor of Electrical and Computer Engineering
North Dakota State University

SYNTHESIS LECTURES ON BIOMEDICAL ENGINEERING #5

ABSTRACT

This is the first in a series of short books on probability theory and random processes for biomedical engineers. This text is written as an introduction to probability theory. The goal was to prepare students, engineers and scientists at all levels of background and experience for the application of this theory to a wide variety of problems—as well as pursue these topics at a more advanced level. The approach is to present a unified treatment of the subject. There are only a few key concepts involved in the basic theory of probability theory. These key concepts are all presented in the first chapter. The second chapter introduces the topic of random variables. Later chapters simply expand upon these key ideas and extend the range of application. A considerable effort has been made to develop the theory in a logical manner—developing special mathematical skills as needed. The mathematical background required of the reader is basic knowledge of differential calculus. Every effort has been made to be consistent with commonly used notation and terminology—both within the engineering community as well as the probability and statistics literature. Biomedical engineering examples are introduced throughout the text and a large number of self-study problems are available for the reader.

KEYWORDS

Probability Theory, Random Processes, Engineering Statistics, Probability and Statistics for Biomedical Engineers, Statistics.

Contents

Preface

This is the first in a series of short books on probability theory and random processes for biomedical engineers. This text is written as an introduction to probability theory. The goal was to prepare students at the sophomore, junior or senior level for the application of this theory to a wide variety of problems—as well as pursue these topics at a more advanced level. Our approach is to present a unified treatment of the subject. There are only a few key concepts involved in the basic theory of probability theory. These key concepts are all presented in the first chapter. The second chapter introduces the topic of random variables. Later chapters simply expand upon these key ideas and extend the range of application.

A considerable effort has been made to develop the theory in a logical manner—developing special mathematical skills as needed. The mathematical background required of the reader is basic knowledge of differential calculus. Every effort has been made to be consistent with commonly used notation and terminology—both within the engineering community as well as the probability and statistics literature.

The applications and examples given reflect the authors' background in teaching probability theory and random processes for many years. We have found it best to introduce this material using simple examples such as dice and cards, rather than more complex biological and biomedical phenomena. However, we do introduce some pertinent biomedical engineering examples throughout the text.

Students in other fields should also find the approach useful. Drill problems, straightforward exercises designed to reinforce concepts and develop problem solution skills, follow most sections. The answers to the drill problems follow the problem statement in random order. At the end of each chapter is a wide selection of problems, ranging from simple to difficult, presented in the same general order as covered in the textbook.

We acknowledge and thank William Pruehsner for the technical illustrations. Many of the examples and end of chapter problems are based on examples from the textbook by Drake [9].

CHAPTER 1

Introduction

We all face uncertainty. A chance encounter, an unpredicted rain or something more serious such as an auto accident or illness. Ordinarily, the uncertainty faced in our daily routine is never quantified and is left as a feeling or intuition. In engineering applications, however, uncertainty must be quantitatively defined, and analyzed in a mathematically rigorous manner resulting in an appropriate and consistent solution. Probability theory provides the tools to analyze, in a deductive manner, the nondeterministic or random aspects of a problem. Our goal is to develop this theory in an axiomatic framework and to demonstrate how it can be used to solve many practical problems in electrical engineering.

In this first chapter, we introduce the elementary aspects of probability theory upon which the following chapter on random variables and chapters in subsequent short books are based. The discussion of probability theory in this book provides a strong foundation for continued study in virtually every field of biomedical engineering, and many of the techniques developed may also be applied to other disciplines.

The theory of probability provides procedures for analyzing random phenomena, phenomena which exhibit behavior that is unpredictable or cannot be determined exactly. Moreover, understanding probability theory is essential before one can use statistics. An easy way to explain what is meant by probability theory is to examine several physical situations that lead to probability theory problems. First consider tossing a fair coin and predicting the outcome of the toss. It is impossible to exactly predict the outcome of the coin flip, so the most we can do is state a chance of our prediction occurring. Next, consider telemetry or a communication system. The signal received consists of the message and/or data plus an undesired signal called thermal noise which is heard as a hiss. The noise is caused by the thermal or random motion of electrons in the conducting media of the receiver—wires, resistors, etc. The signal received also contains noise picked up as the signal travels through the atmosphere. Note that it is impossible to exactly compute the value of the noise caused by the random motion of the billions of charged particles in the receiver's amplification stages or added in the environment. Thus, it is impossible to completely remove the undesired noise from the signal. We will see, however, that probability theory provides a means by which most of the unwanted noise is removed.

From the previous discussion, one might argue that our inability to exactly compute the value of thermal noise at every instant of time is due to our ignorance, and that if a better model of this phenomenon existed, then thermal noise could be exactly described. Actually, thermal noise is well understood through extensive theoretical and experimental studies, and exactly characterizing it would be at least as difficult as trying to exactly predict the outcome of a fair coin toss: the process is inherently indeterminant.

On the other hand, one can take the point of view that one is really interested in the average behavior of certain complicated processes—such as the average error rate of a communication system or the efficacy of a drug treatment program. Probability theory provides a useful tool for studying such problems even when one could argue whether or not the underlying phenomenon is truly "random."

Other examples of probability theory used in biomedical engineering include:

- Diffusion of ions across a cell membrane [3, 15]
- Biochemical reactions [15]
- Muscle model using the cross-bridge model for contraction [15]
- Variability seen in the genetic makeup of a species as DNA is transferred from one generation to another. That is, developing a mathematical model of DNA mutation processes and reconstruction of evolutionary relationships between modern species [3, 20].
- Genetics [3]
- Medical tests [26]
- Infectious diseases [2, 3, 14]
- Neuron models and synaptic transmission [15]
- Biostatistics [26]

Because the complexity of the previous biomedical engineering models obscures the application of probability theory, most of the examples presented are straightforward applications involving cards and dice. After a concept is presented, some biomedical engineering examples are introduced.

We begin with some preliminary concepts necessary for our study of probability theory. Students familiar with set theory and the mathematics of counting (permutations and combinations) should find it rapid reading, however, it should be carefully read by everyone. After these preliminary concepts have been covered, we then turn our attention to the axiomatic development of probability theory.

1.1 PRELIMINARY CONCEPTS

We begin with a discussion of set theory in order to establish a common language and notation. While much of this material is already familiar to you, we also want to review the basic set operations which are important in probability theory. As we will see, the definitions and concepts presented here will clarify and unify the mathematical foundations of probability theory. The following definitions and operations form the basics of set theory.

Definition 1.1.1. *A **set** is an unordered collection of objects. We typically use a capital letter to denote a set, listing the objects within braces or by graphing. The notation $A = \{x : x > 0, x \leq 2\}$ is read as "the set A contains all x such that x is greater than zero and less than or equal to two." The notation $\zeta \in A$ is read as "the object zeta is in the set A." Two sets are equal if they have exactly the same objects in them; i.e., $A = B$ if A contains exactly the same elements that are contained in B.*

*The **null set**, denoted $\varnothing$, is the empty set and contains no objects.*

*The **universal set**, denoted S, is the set of all objects in the universe. The universe can be anything we define it to be. For example, we sometimes consider $S = \mathbb{R}$, the set of all real numbers.*

If every object in set A is also an object in set B, then A is a subset of B. We shall use the notation $A \subset B$ to indicate that A is a subset of B. The expression $B \supset A$ (read as " A contains B ") is equivalent to $A \subset B$.

*The **union** of sets A and B, denoted $A \cup B$, is the set of objects that belong to A or B or both; i.e., $A \cup B = \{\zeta : \zeta \in A \text{ or } \zeta \in B\}$.*

*The **intersection** of sets A and B, denoted $A \cap B$, is the set of objects common to both A and B; i.e., $A \cap B = \{\zeta : \zeta \in A \text{ and } \zeta \in B\}$.*

*The **complement** of a set A, denoted A^c, is the collection of all objects in S not included in A; i.e., $A^c = \{\zeta \in S : \zeta \notin A\}$.*

These definitions and relationships among sets are illustrated in Fig. 1.1. Such diagrams are called Venn diagrams. Sets are represented by simple plane areas within the universal set, pictured as a rectangle. Venn diagrams are important visual aids which may help us to understand relationships among sets; however, proofs must be based on definitions and theorems. For example, the above definitions can be used to show that if $A \subset B$ and $B \subset A$ then $A = B$; this fact can then be used whenever it is necessary to show that two sets are equal.

Theorem 1.1.1. *Let $A \subset B$ and $B \subset A$. Then $A = B$.*

Proof. We first note that the empty set is a subset of any set. If $A = \varnothing$ then $B \subset A$ implies that $B = \varnothing$. Similarly, if $B = \varnothing$ then $A \subset B$ implies that $A = \varnothing$.

The theorem is obviously true if A and B are both empty.

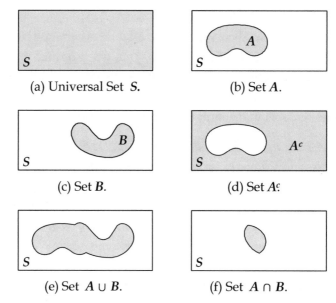

(a) Universal Set *S*.

(b) Set *A*.

(c) Set *B*.

(d) Set *A*ᶜ

(e) Set *A* ∪ *B*.

(f) Set *A* ∩ *B*.

FIGURE 1.1: Venn diagrams

Assume $A \subset B$ and $B \subset A$, and that A and B are nonempty. Since $A \subset B$, if $\zeta \in A$ then $\zeta \in B$. Since $B \subset A$, if $\zeta \in B$ then $\zeta \in A$. We conclude that $A = B$.   ∎

The converse of the above theorem is also true: If $A = B$ then $A \subset B$ and $B \subset A$.

Whereas a set is an unordered collection of objects, a set can be an unordered collection of ordered objects. The following examples illustrate common ways of specifying sets of two-dimensional real numbers.

Example 1.1.1. *Let* $A = \{(x, y) : y - x = 1\}$ *and* $B = \{(x, y) : x + y = 1\}$. *Find the set* $A \cap B$. *The notation* (x, y) *denotes an ordered pair.*

Solution. A pair $(x, y) \in A \cap B$ only if $y = 1 + x$ and $y = 1 - x$; consequently, $x = 0$, $y = 1$, and

$$A \cap B = \{(x, y) : x = 0, y = 1\}.$$ ∎

Example 1.1.2. *Let* $A = \{(x, y) : y \le x\}$, $B = \{(x, y) : x \le y + 1\}$, $C = \{(x, y) : y < 1\}$, *and* $D = \{(x, y) : 0 \le y\}$. *Find and sketch* $E = A \cap B$, $F = C \cap D$, $G = E \cap F$, *and* $H = \{(x, y) : (-x, y + 1) \in G\}$.

Solution. The solutions are easily found with the aid of a few quick sketches. First, sketch the boundaries of the given sets A, B, C, and D. If the boundary of the region is included in the set, it is indicated with a solid line. If the "boundary" is not included, it is indicated with a dotted line in the sketch.

We have

$$E = A \cap B = \{(x, y) : x - 1 \leq y \leq x\}$$

and

$$F = C \cap D = \{(x, y) : 0 \leq y < 1\}.$$

The set G is the set of all ordered pairs (x, y) satisfying both $x - 1 \leq y \leq x$ and $0 \leq y < 1$. Using 1^- to denote a value just less than 1, the second inequality may be expressed as $0 \leq y \leq 1^-$. We may then express the set G as

$$G = \{(x, y) : \max\{0, x - 1\} \leq y \leq \min\{x, 1^-\}\},$$

where $\max\{a, b\}$ denotes the maximum of a and b; similarly, $\min\{a, b\}$ denotes the minimum of a and b.

The set H is obtained from G by folding about the y-axis and translating down one unit. This can be seen from the definitions of G and H by noting that $(x, y) \in H$ if $(-x, y + 1) \in G$; hence, we replace x with $-x$ and y with $y + 1$ in the above result for G to obtain

$$H = \{(x, y) : \max\{0, -x - 1\} \leq y + 1 \leq \min\{-x, 1^-\}\},$$

or

$$H = \{(x, y) : \max\{-1, -x - 2\} \leq y \leq \min\{-1 - x, 0^-\}\}.$$

The sets are illustrated in Fig. 1.2. ■

1.1.1 Operations on Sets

Throughout probability theory it is often required to establish relationships between sets. The set operations $\cup$ and $\cap$ operate on sets in much the same way the operations $+$ and $\times$ operate on real numbers. Similarly, the special sets $\varnothing$ and S correspond to the additive identity 0 and the multiplicative identity 1, respectively. This correspondence between operations on sets and operations on real numbers is made explicit by the theorem below, which can be proved by applying the definitions of the basic set operations stated above. The reader is strongly encouraged to carry out the proof.

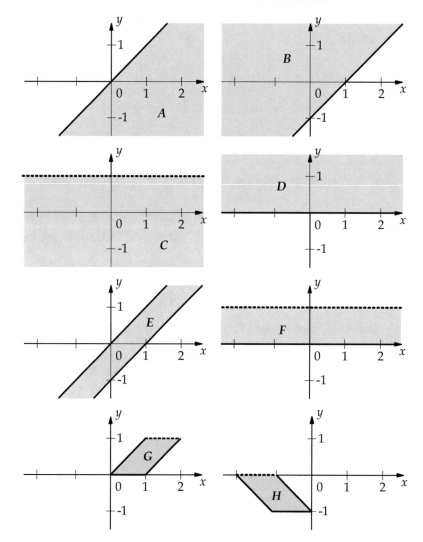

FIGURE 1.2: Sets for Example 1.1.2

Theorem 1.1.2 (Properties of Set Operations). *Let A, B, and C be subsets of S. Then*
    Commutative Properties

$$A \cup B = B \cup A \qquad (1.1)$$
$$A \cap B = B \cap A \qquad (1.2)$$

Associative Properties

$$A \cup (B \cup C) = (A \cup B) \cup C \qquad (1.3)$$
$$A \cap (B \cap C) = (A \cap B) \cap C \qquad (1.4)$$

Distributive Properties

$$A \cap (B \cup C) = (A \cap B) \cup (A \cap C) \qquad (1.5)$$
$$A \cup (B \cap C) = (A \cup B) \cap (A \cup C) \qquad (1.6)$$

De Morgan's Laws

$$(A \cap B)^c = A^c \cup B^c \qquad (1.7)$$
$$(A \cup B)^c = A^c \cap B^c \qquad (1.8)$$

Identities involving ∅ and S

$$A \cup \emptyset = A \qquad (1.9)$$
$$A \cap S = A \qquad (1.10)$$
$$A \cap \emptyset = \emptyset \qquad (1.11)$$
$$A \cup S = S \qquad (1.12)$$

Identities involving complementation

$$A \cap A^c = \emptyset \qquad (1.13)$$
$$A \cup A^c = S \qquad (1.14)$$
$$(A^c)^c = A \qquad (1.15)$$

Additional insight to operations on sets is provided by the correspondence between the algebra of set inclusion and Boolean algebra. An element either belongs to a set or it does not. Thus, interpreting sets as Boolean (logical) variables having values of 0 or 1, the $\cup$ operation as the logical OR, the $\cap$ as the logical AND operation, and the ^{c} as the logical complement, any expression involving set operations can be treated as a Boolean expression.

The following two theorems provide additional tools to apply when solving problems involving set operations. The Principle of Duality reveals that about half of the set identities in the above theorem are redundant. Note that a set identity is an expression which remains true for arbitrary sets. The dual of a set identity is also a set identity. The dual of an arbitrary set expression is *not* in general the same as the original expression.

Theorem 1.1.3 (Negative Absorption Theorem)

$$A \cup (A^c \cap B) = A \cup B. \qquad (1.16)$$

Proof. Using the distributive property,

$$A \cup (A^c \cap B) = (A \cup A^c) \cap (A \cup B)$$
$$= S \cap (A \cup B)$$
$$= A \cup B. \qquad \blacksquare$$

Theorem 1.1.4 (Principle of Duality). *Any set identity remains true if the symbols*

$$\cup, \cap, S, \text{ and } \varnothing$$

are replaced with the symbols

$$\cap, \cup, \varnothing, \text{ and } S,$$

respectively.

Proof. The proof follows by applying De Morgan's Laws and renaming sets A^c, B^c, etc. as A, B, etc. $\qquad \blacksquare$

Example 1.1.3. *Verify the following set identity:*

$$A \cup (B^c \cup ((A^c \cup B) \cap C))^c = A \cup (B \cap C^c).$$

Solution. From the duality principle, the given expression is equivalent to

$$A \cap (B^c \cap ((A^c \cap B) \cup C))^c = A \cap (B \cup C^c).$$

Using the distributive property and applying De Morgan's Law we obtain

$$(B^c \cap ((A^c \cap B) \cup C))^c = ((B^c \cap A^c \cap B) \cup (B^c \cap C))^c$$
$$= ((B^c \cap B \cap A^c) \cup (B^c \cap C))^c$$
$$= ((\varnothing \cap A^c) \cup (B^c \cap C))^c$$
$$= (B^c \cap C)^c$$
$$= B \cup C^c,$$

from which the desired result follows.

Of course, there are always alternatives for problem solutions. For this example, one could begin by applying the distributive property as follows:

$$(B^c \cup ((A^c \cup B) \cap C))^c = ((B^c \cup A^c \cup B) \cap (B^c \cup C))^c$$
$$= ((B^c \cup B \cup A^c) \cap (B^c \cup C))^c$$
$$= ((S \cup A^c) \cap (B^c \cup C))^c$$
$$= (B^c \cup C)^c$$
$$= B \cap C^c.$$

Theorem 1.1.2 is easily extended to deal with any finite number of sets. To do this, we need notation for the union and intersection of a collection of sets.

Definition 1.1.2. *We define the **union** of a collection of sets (one can refer to such a collection of sets as a "set of sets")*

$$\{A_i : i \in I\} \tag{1.17}$$

by

$$\bigcup_{i \in I} = \{\zeta \in S : \zeta \in A_i \text{ for some } i \in I\} \tag{1.18}$$

*and the **intersection** of a collection of sets*

$$\{A_i : i \in I\} \tag{1.19}$$

by

$$\bigcap_{i \in I} A_i = \{\zeta \in S : \zeta \in A_i \text{ for every } i \in I\}. \tag{1.20}$$

We note that if $I = \varnothing$ then

$$\bigcup_{i \in I} A_i = \varnothing \tag{1.21}$$

and

$$\bigcap_{i \in I} A_i = S. \tag{1.22}$$

For example, if $I = \{1, 2, \ldots, n\}$, then we have

$$\bigcup_{i \in I} A_i = \bigcup_{i=1}^{n} A_i = \begin{cases} A_i \cup A_2 \cup \cdots \cup A_n, & \text{if } n \geq 1 \\ \varnothing, & \text{if } n < 1, \end{cases} \tag{1.23}$$

and

$$\bigcap_{i \in I} A_i = \bigcap_{i=1}^{n} A_i = \begin{cases} A_i \cap A_2 \cap \cdots \cap A_n, & \text{if } n \geq 1 \\ S, & \text{if } n < 1. \end{cases} \tag{1.24}$$

Theorem 1.1.5 (Properties of Set Operations). *Let $A_1, A_2, \ldots, A_n$, and B be subsets of S. Then*

Commutative and Associative Properties

$$\bigcup_{i=1}^{n} A_i = A_1 \cup A_2 \cup \cdots \cup A_n = A_{i_1} \cup A_{i_2} \cup \cdots \cup A_{i_n}, \tag{1.25}$$

and

$$\bigcap_{i=1}^{n} A_i = A_1 \cap A_2 \cap \cdots \cap A_n = A_{i_1} \cap A_{i_2} \cap \cdots \cap A_{i_n}, \qquad (1.26)$$

where $i_1 \in \{1, 2, \ldots, n\} = I_1$, $i_2 \in I_2 = I_1 \cap \{i_1\}^c$, and

$$i_\ell \in I_\ell = I_{\ell-1} \cap \{i_{\ell-1}\}^c, \quad \ell = 2, 3, \ldots, n.$$

In other words, the union (or intersection) of n sets is independent of the order in which the unions (or intersections) are taken.

Distributive Properties

$$B \cap \bigcup_{i=1}^{n} A_i = \bigcup_{i=1}^{n} (B \cap A_i) \qquad (1.27)$$

$$B \cup \bigcap_{i=1}^{n} A_i = \bigcap_{i=1}^{n} (B \cup A_i) \qquad (1.28)$$

De Morgan's Laws

$$\left(\bigcap_{i=1}^{n} A_i \right)^c = \bigcup_{i=1}^{n} A_i^c \qquad (1.29)$$

$$\left(\bigcup_{i=1}^{n} A_i \right)^c = \bigcap_{i=1}^{n} A_i^c \qquad (1.30)$$

Throughout much of probability, it is useful to decompose a set into a union of simpler, non-overlapping sets. This is an application of the "divide and conquer" approach to problem solving. Necessary terminology is established in the following definition.

Definition 1.1.3. *The sets $A_1, A_2, \ldots, A_n$ are **mutually exclusive** (or **disjoint**) if*

$$A_i \cap A_j = \varnothing$$

*for all i and j with $i \neq j$. The sets $A_1, A_2, \ldots, A_n$ form a **partition** of the set B if they are mutually exclusive and*

$$B = A_1 \cup A_2 \cup \cdots \cup A_n = \bigcup_{i=1}^{\cup} A_i$$

*The sets $A_1, A_2, \ldots, A_n$ are **collectively exhaustive** if*

$$S = A_1 \cup A_2 \cup \cdots \cup A_n = \bigcup_{i=1}^{n} A_i.$$

Example 1.1.4. *Let $S=\{(x, y) : x \geq 0, y \geq 0\}$, $A=\{(x, y) : x + y < 1\}$, $B = \{(x, y) : x < y\}$, and $C = \{(x, y) : xy > 1/4\}$. Are the sets A, B, and C mutually exclusive, collectively exhaustive, and/or a partition of S?*

Solution. Since $A \cap C = \varnothing$, the sets A and C are mutually exclusive; however, $A \cap B \neq \varnothing$ and $B \cap C \neq \varnothing$, so A and B, and B and C are not mutually exclusive. Since $A \cup B \cup C \neq S$, the events are not collectively exhaustive. The events A, B, and C are not a partition of S since they are not mutually exclusive and collectively exhaustive. ∎

Definition 1.1.4. *The **Cartesian product** of sets A and B is a set of ordered pairs of elements of A and the elements of B:*

$$A \times B = \{\zeta = (\zeta_1, \zeta_2) : \zeta_1 \in A, \zeta_2 \in B\}. \tag{1.31}$$

The Cartesian product of sets $A_1, A_2, \ldots, A_n$ is a set of n-tuples (an ordered list of n elements) of elements of $A_1, A_2, \ldots, A_n$:

$$A_1 \times A_2 \times \cdots \times A_n = \{\zeta = (\zeta_1, \zeta_2, \ldots \zeta_n) : \zeta_1 \in A_1, \zeta_2 \in A_2, \ldots, \zeta_n \in A_n\}. \tag{1.32}$$

An important example of a Cartesian product is the usual n-dimensional real Euclidean space:

$$R^n = \underbrace{R \times R \times \cdots \times R}_{n \text{ terms}}. \tag{1.33}$$

1.1.2 Notation
We briefly present a collection of some frequently used (and confused) notation.

Some special sets of real numbers will often be encountered:

$$(a, b) = \{x : a < x < b\},$$

$$(a, b] = \{x : a < x \leq b\},$$

$$[a, b) = \{x : a \leq x < b\},$$

and

$$[a, b] = \{x : a \leq x \leq b\}.$$

Note that if $a > b$, then $(a, b) = (a, b] = [a, b) = [a, b] = \varnothing$. If $a = b$, then $(a, b) = (a, b] = [a, b) = \varnothing$ and $[a, b] = a$. The notation (a, b) is also used to denote an ordered pair—we depend on the context to determine whether (a, b) represents an open interval of real numbers or an ordered pair.

We will often encounter unions and intersections of a collection of indexed sets. The shorthand notations

$$\bigcup_{i=m}^{n} A_i = \begin{cases} A_m \cup A_{m+1} \cup \cdots \cup A_n, & \text{if } n \geq m \\ \varnothing, & \text{if } n < m, \end{cases} \qquad (1.34)$$

and

$$\bigcap_{i=m}^{n} A_i = \begin{cases} A_m \cap A_{m+1} \cap \cdots \cap A_n, & \text{if } n \geq m \\ S, & \text{if } n < m \end{cases} \qquad (1.35)$$

are useful for reducing the length of expressions. These conventions are similar to the notation used to express sums and products of real numbers:

$$\sum_{i=m}^{n} x_i = \begin{cases} x_m + x_{m+1} + \cdots + x_n, & \text{if } n \geq m \\ 0, & \text{if } n < m, \end{cases} \qquad (1.36)$$

and

$$\prod_{i=m}^{n} x_i = \begin{cases} x_m \times x_{m+1} + \cdots + x_n, & \text{if } n \geq m \\ 1, & \text{if } n < m. \end{cases} \qquad (1.37)$$

As with integration, a change of variable is often helpful in solving problems and proving theorems. Consider, for example, using the change of summation index $j = n - i$ to obtain

$$\sum_{i=1}^{n} x_i = \sum_{j=0}^{n-1} x_{n-j}.$$

A corresponding change of integration variable $\lambda = t - \tau$ yields

$$\int_{0}^{t} f(\tau)d\tau = -\int_{t}^{0} f(t - \lambda)d\lambda = \int_{0}^{t} f(t - \lambda)d\lambda.$$

Note that

$$\sum_{i=1}^{3} i = 6 \neq -\sum_{i=3}^{1} i = 0,$$

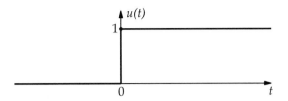

FIGURE 1.3: Unit-step function

whereas

$$\int_1^3 x\,dx = 4 = -\int_3^1 x\,dx.$$

In addition to the usual trigonometric functions $\sin(\cdot)$, $\cos(\cdot)$, and exponential function $\exp(x) = e^x$, we will make use of the **unit step function** $u(t)$ defined as

$$u(t) = \begin{cases} 1, & \text{if } t \geq 0 \\ 0, & \text{if } t < 0. \end{cases} \qquad (1.38)$$

In particular, we define $u(0) = 1$, which proves to be convenient for our discussions of distribution functions in Chapter 2. The unit step function is illustrated in Fig. 1.3.

Drill Problem 1.1.1. *Define the sets* $S = \{1, 2, \ldots, 9\}$, $A = \{1, 2, 3, 4\}$, $B = \{4, 5, 8\}$, *and* $C = \{3, 4, 7, 8\}$. *Determine the sets:* $(a)\ (A \cap B)^c$, $(b)\ (A \cup B \cup C)^c$, $(c)\ (A \cap B) \cup (A \cap B^c) \cup (A^c \cap B)$, $(d)\ A \cup (A^c \cap B) \cup ((A \cup (A^c \cap B))^c \cap C)$.

Answers: $\{1, 2, 3, 4, 5, 7, 8\}$; $\{6, 9\}$; $\{1, 2, 3, 5, 6, 7, 8, 9\}$; $\{1, 2, 3, 4, 5, 8\}$.

Drill Problem 1.1.2. *Using the properties of set operations (not Venn diagrams), determine the validity of the following relationships for arbitrary sets* A, B, C, *and* D: $(a)\ B \cup (B^c \cap A) = A \cup B$, $(b)\ (A \cap B) \cup (A \cap B^c) \cup (A^c \cap B) = A$, $(c)\ C \cap (A \cup ((A^c \cup B^c)^c \cap D)) = A \cap C$, $(d)\ A \cup (A^c \cup B^c)^c = A$.

Answers: True, True, True, False.

1.2 THE SAMPLE SPACE

An experiment is a model of a random phenomenon, an abstraction which ignores many of the dynamic relationships of the actual random phenomenon. We seek to capture only the prominent features of the real world problem with our experiment so that needless details do not obscure our analysis. Consider our model of resistance, $v(t) = Ri(t)$. One can utilize more accurate models of resistance that will improve the accuracy of our real world analysis, but the cost is far too great to sacrifice the simplicity of $v(t) = Ri(t)$ for our work with circuit analysis problems.

Now, we will associate the universal set S with the set of outcomes of an experiment describing a random phenomenon. Specifically, the sample space, or outcome space, is the finest grain, mutually exclusive, and collectively exhaustive listing of all possible outcomes for the experiment. Tossing a fair die is an example of an experiment. In performing this experiment, the outcome is the number on the upturned face of the die, and thus the sample space is $S = \{1, 2, 3, 4, 5, 6\}$. Notice that we could have included the distance of the toss, the number of rolls, and other details in addition to the number on the upturned face of the die for the experiment, but unless our analysis specifically called for these details, it would be unreasonable to include them.

A sample space is classified as being **discrete** if it contains a countable number of objects. A set is **countable** if the elements can be placed in one-to-one correspondence with the positive integers. The set of integers $S = \{1, 2, 3, 4, 5, 6\}$ from the die toss experiment is an example of a discrete sample space, as is the set of all integers. In contrast, the set of all real numbers between 0 and 1 is an example of an **uncountable** sample space. For now, we shall be content to deal with discrete outcome spaces. We will later find that probability theory is concerned with another discrete space, called the event space, which is a countable collection of subsets of the outcome space—whether or not the outcome space itself is discrete.

1.2.1   Tree Diagrams

Many experiments consist of a sequence of simpler "subexperiments" as, for example, the sequential tossing of a coin and the sequential drawing of cards from a deck. A **tree diagram** is a useful graphical representation of a sequence of experiments—particularly when each subexperiment has a small number of possible outcomes.

Example 1.2.1.   *A coin is tossed twice. Illustrate the sample space with a tree diagram.*

Solution.   Let H_i denote the outcome of a head on the ith toss and T_i denote the outcome of a tail on the ith toss of the coin. The tree diagram illustrating the sample space for this sequence of two coin tosses is shown in Fig. 1.4. We draw the tree diagram as a combined experiment, in a left to right path from the origin, consisting of the first coin toss (with each of its outcomes) immediately followed by the second coin toss (with each of its outcomes). Note that the combined experiment is really a sequence of two experiments. Each node represents an outcome of one coin toss and the branches of the tree connect the nodes. The number of branches to the right of each node corresponds to the number of outcomes for the next coin toss (or experiment). A sequence of samples connected by branches in a left to right path from the origin to a terminal node represents a sample point for the combined experiment. There

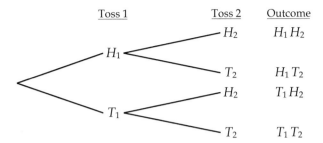

FIGURE 1.4: Tree diagram for Example 1.2.1

is a one-to-one correspondence between the paths in the tree diagram and the sample points in the sample space for the combined experiment. For this example, the outcome space for the combined experiment is

$$S = \{H_1 H_2,\ H_1 T_2,\ T_1 H_2,\ T_1 T_2\},$$

consisting of four sample points. ∎

Example 1.2.2. *Two balls are selected, one after the other, from an urn that contains nine red, five blue, and two white balls. The first ball is not replaced in the urn before the next ball is chosen. Set up a tree diagram to describe the color composition of the sample space.*

Solution. Let R_i, B_i, and W_i, respectively, denote a red, blue, and white ball drawn on the ith draw. Note that R_1 denotes the collection of nine outcomes for the first draw resulting in a red ball. We will refer to such a collection of outcomes as an event. The tree diagram shown in Fig. 1.5 is considerably simplified by using only one branch to represent R_1, instead of nine. Note that if there were only one white ball (instead of two), the branch terminating with the sequence event $W_1 W_2$ would be removed from the tree. ∎

Although the tree diagram seems to imply that each outcome is physically followed by the next one, and so on, this need not be the case. A tree diagram can often be used even if the subexperiments are performed at the same time. A sequential drawing of the sample space is only a convenient representation, not necessarily a physical representation.

Example 1.2.3.[1] *Consider a population with genotypes for the blood disease sickle cell anemia, with the two genotypes denoted by* **a** *and* **b**. *Assume that* **b** *is the code for the anemic trait and that* **a** *is without the trait. Individuals can have the genotypes* **aa**, **ab** *and* **bb**. *Note that* **ab** *and* **ba** *are indistinguishable, and that the disease is present only with* **bb**. *Construct a tree diagram if both parents are* **ab**.

---

[1]This example is based on [14], pages 45–47.

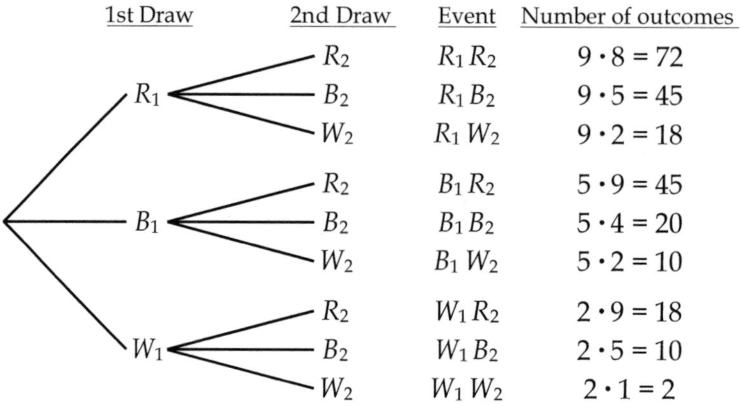

1st Draw	2nd Draw	Event	Number of outcomes
	R_2	$R_1 R_2$	$9 \cdot 8 = 72$
R_1	B_2	$R_1 B_2$	$9 \cdot 5 = 45$
	W_2	$R_1 W_2$	$9 \cdot 2 = 18$
	R_2	$B_1 R_2$	$5 \cdot 9 = 45$
B_1	B_2	$B_1 B_2$	$5 \cdot 4 = 20$
	W_2	$B_1 W_2$	$5 \cdot 2 = 10$
	R_2	$W_1 R_2$	$2 \cdot 9 = 18$
W_1	B_2	$W_1 B_2$	$2 \cdot 5 = 10$
	W_2	$W_1 W_2$	$2 \cdot 1 = 2$

FIGURE 1.5: Tree diagram for Example 1.2.2

Solution. Background for this problem is given in footnote.[2] The tree diagram shown in Fig. 1.6 is formed by first listing the alleles of the first parent and then the alleles of the second parent. Notice only one child out of four will have sickle cell anemia.

[2]Genetics is the study of the variation within a species, originally based on the work by Mendel in the 19th century. Reproduction is based on transferring genetic information from one generation to the next. Mendel originally called genetic information *traits* in his work with peas (i.e., stem length characteristic as tall and dwarf, seed shape characteristic as round or wrinkled, etc.). Today we refer to traits as genes, with variations in genes called *alleles* or *genotypes*. Each parent stores two genes for each characteristic, and passes only one gene to the progeny. Each gene is equally likely of being passed by the parent to the progeny. Through breeding, Mendel was able to create pure strains of the pea plant, strains that produced only one type of progeny that was identical to the parents (i.e., the two genes were identical). By studying one characteristic at a time, Mendel was able to examine the impact of pure parent traits on the progeny. In the progeny, Mendal discovered that one trait was dominant and the other recessive or hidden. The dominant trait was observed when either parent passed a dominant trait. The recessive trait was observed when both parents passed the recessive trait. Mendel showed that when both parents displayed the dominant trait, offspring could be produced with the recessive trait if both parents contained a dominant and recessive trait, and both passed the recessive trait.

Genetic information is stored in DNA. DNA is a double helix, twisted ladder-like molecule, where pairs of nucleotides appear at each rung joined by a hydrogen bond. The nucleotides are called adenine (A), guanine (G), cytosine (C) and thymine (T). Each nucleotide has a complementary nucleotide that forms a rung; A is always paired with T and G with C, and is directional. Thus if one knows one chain, the other is known. For instance, if one is given AGGTCT, the complement is TCCAGA.

DNA is also described by nucleosomes that are organized into pairs of chromosomes. Chromosomes store all information about the organism's chemical needs and information about inheritable traits. Humans contain 23 matched pairs of chromosomes. Each chromosome contains thousands of genes that encode instructions for the manufacture of proteins (actually, this process is carried out by messenger RNA)—they are the blueprint for the individual. Each gene has a particular location in a specific chromosome. Slight gene variations exist within a population.

DNA replication occurs during cell division where the double helix is unzipped by an enzyme that breaks the hydrogen bonds that form the ladder rungs, leaving two strands. New double strands are then formed by an elaborate error checking process that binds the appropriate complementary nucleotides. While this process involves

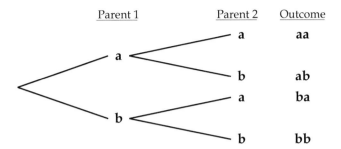

FIGURE 1.6: Tree diagram for Example 1.2.3

1.2.2 Coordinate System

Another approach to illustrating the outcome space is to use a **coordinate system** representation. This approach is especially useful when the combined experiment involves a combination of two experiments with numerical outcomes. With this method, each axis lists the outcomes for each subexperiment.

Example 1.2.4. *A die is tossed twice. Illustrate the sample space using a coordinate system.*

Solution. The coordinate system representation is shown in Fig. 1.7. Note that there are 36 sample points in the experiment; six possible outcomes on the first toss of the die times six possible outcomes on the second toss of the die, each of which is indicated by a point in the coordinate space. Additionally, we distinguish between sample points with regard to order; e.g., (1,2) is different from (2,1). ∎

Example 1.2.5. *A real number x is chosen "at random" from the interval $[0, 10]$. A second real number y is chosen "at random" from the interval $[0, x]$. Illustrate the sample space using a coordinate system.*

Solution. The coordinate system representation is shown in Fig. 1.8. Note that there are an uncountable number of sample points in this experiment. ∎

1.2.3 Mathematics of Counting

Although either a tree diagram or a coordinate system enables us to determine the number of outcomes in the sample space, problems immediately arise when the number of outcomes is large. To easily and efficiently solve problems in many probability theory applications, it is

---

minimal errors (approx. one per billion), errors do happen. The most common error is nucleotide substitution where one is changed for another. For instance, AGGTCT becomes AGCTCT (i.e., the third site goes from G to C). Additional information on this topic is found in [3, 10, 14].

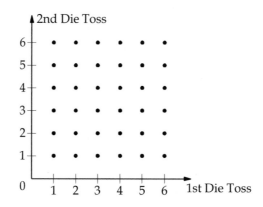

FIGURE 1.7: Coordinate system outcome space for Example 1.2.4

important to know the number of outcomes as well as the number of subsets of outcomes with a specified composition. We now develop some formulas which enable us to count the number of outcomes without the aid of a tree diagram or a coordinate system. These formulas are a part of a branch of mathematics known as combinatorial analysis.

Sequence of Experiments

Suppose a combined experiment is performed in which the first experiment has n_1 possible outcomes, followed by a second experiment which has n_2 possible outcomes, followed by a third experiment which has n_3 possible outcomes, etc. A sequence of k such experiments thus has

$$n = n_1 n_2 \cdots n_k \tag{1.39}$$

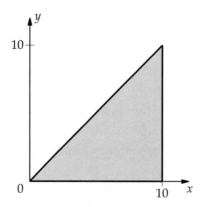

FIGURE 1.8: Coordinate system outcome space for Example 1.2.5

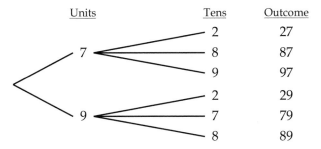

Units	Tens	Outcome
	2	27
7	8	87
	9	97
	2	29
9	7	79
	8	89

FIGURE 1.9: Tree diagram for Example 1.2.6

possible outcomes. This result allows us to quickly calculate the number of sample points in a sequence of experiments without drawing a tree diagram, although visualizing the tree diagram will lead instantly to the above equation.

Example 1.2.6. *How many odd two digit numbers can be formed from the digits 2, 7, 8, and 9, if each digit can be used only once?*

Solution. A tree diagram for this sequential drawing of digits is shown in Fig. 1.9. From the origin, there are two ways of selecting a number for the unit's place (the first experiment). From each of the nodes in the first experiment, there are three ways of selecting a number for the ten's place (the second experiment). The number of outcomes in the combined experiment is the product of the number of branches for each experiment, or $2 \times 3 = 6$. ∎

Example 1.2.7. *An analog-to-digital (A/D) converter outputs an 8-bit word to represent an input analog voltage in the range −5 to +5 V. Determine the total number of words possible and the maximum sampling (quantization) error.*

Solution. Since each bit (or binary digit) in a computer word is either a one or a zero, and there are 8 bits, then the total number of computer words is

$$n = 2^8 = 256.$$

To determine the maximum sampling error, first compute the range of voltage assigned to each computer word which equals

$$10 \text{ V}/256 \text{ words} = 0.0390625 \text{ V/word}$$

and then divide by two (i.e., round off to the nearest level), which yields a maximum error of 0.0195312 V/word. ∎

Sampling With Replacement

Definition 1.2.1. *Let $S_n = \{\zeta_1, \zeta_2, \ldots, \zeta_n\}$; i.e., S_n is a set of n arbitrary objects. A **sample of size** k **with replacement** from S_n is an ordered list of k elements:*

$$(\zeta_{i_1}, \zeta_{i_2}, \ldots, \zeta_{i_k}),$$

where $i_j \in \{1, 2, \ldots, n\}$, for $j = 1, 2, \ldots, k$.

Theorem 1.2.1. *There are n^k samples of size k when sampling with replacement from a set of n objects.*

Proof. Let S_n be the set of n objects. Each component of the sample can have any of the n values contained in S_n, and there are k components, so that there are n^k distinct samples of size k (with replacement) from S_n. ∎

Example 1.2.8. *An urn contains ten different balls B_1, B_2, \ldots, B_{10}. If k draws are made from the urn, each time replacing the ball, how many samples are there?*

Solution. There are 10^k such samples of size k. ∎

Example 1.2.9. *How many k-digit base b numbers are there?*

Solution. There are b^k different base b numbers. ∎

Permutations

Definition 1.2.2. *Let $S_n = \{\zeta_1, \zeta_2, \ldots, \zeta_n\}$; i.e., S_n is a set of n arbitrary objects. A **sample of size** k **without replacement** or **permutation** from S_n is an ordered list of k elements of S_n:*

$$(\zeta_{i_1}, \zeta_{i_2}, \ldots, \zeta_{i_k}),$$

where $i_j \in I_{n,j}$, $I_{n,1} = \{1, 2, \ldots, n\}$, and $I_{n,j} = I_{n,j-1} \cap \{i_{j-1}\}^c$, for $j = 2, 3, \ldots, k$.

Theorem 1.2.2. *There are*

$$P_{n,k} = n(n-1)\cdots(n-k+1)$$

*distinct samples of size k without replacement from a set of n objects. The quantity $P_{n,k}$ is also called the **number of permutations of** n **things taken** k **at a time** and can be expressed as*

$$P_{n,k} = \frac{n!}{(n-k)!}, \tag{1.40}$$

where $n! = n(n-1)(n-2)\cdots 1$, and $0! = 1$.

Proof. We note that the first component of the sample can have any of n values, the second can have any of $n-1$ values, the jth component can have any of $n-(j-1)$ values. Consequently,

$$P_{n,k} = n(n-1)\cdots(n-k+1). \qquad \blacksquare$$

Example 1.2.10. *From a rural community of 40 people, four people are selected to serve on a committee. Those selected are to serve as president, vice president, treasurer, and secretary. Find the number of sample points in* S.

Solution. Since the order of selection is important, we compute the number of sample points in S using the formula for permutations with $n = 40$ and $k = 4$ as

$$P_{40,4} = \frac{40!}{(40-4)!} = 40 \times 39 \times 38 \times 37 = 2{,}193{,}360. \qquad \blacksquare$$

It is important to emphasize that sampling either with or without replacement yields an *ordered* list: samples consisting of the same elements occurring in a different order are counted as distinct samples.

Next, consider a case in which some of the n objects are identical and indistinguishable.

Theorem 1.2.3. *The number of distinct permutations of n objects taken n at a time in which n_1 are of one kind, n_2 are of a second kind, $\ldots$, n_k are of a kth kind, is*

$$P_{n:n_1,n_2,\ldots,n_k} = \frac{n!}{n_1!\, n_2! \cdots n_k!}, \qquad (1.41)$$

where

$$n = \sum_{i=1}^{k} n_i. \qquad (1.42)$$

Proof. This result can be verified by noting that there are $n!$ ways of ordering n things ($n!$ samples without replacement from n things). We must divide this by $n_1!$ (the number of ways of ordering n_1 things), then by $n_2!$, etc. For example, if $A = \{a_1, a_2, a_3, b\}$, then the number of permutations taken 4 at a time is 4!. If the subscripts are disregarded, then $a_1a_2ba_3$ is identical to $a_1a_3ba_2$, $a_2a_3ba_1$, $a_3a_1ba_2$, $a_2a_1ba_3$ and $a_3a_2ba_1$, and can not be included as unique permutations. In this example then, the total number of permutations (4!) is divided by the number of permutations of the three As and equals 4!/3!. Thus, whenever a number of identical objects form part of a sample, the number of total permutations of all the objects is divided by the product of the number of permutations due to each of the identical objects. $\qquad \blacksquare$

Example 1.2.11. *How many different 8-bit computer words can be formed from five zeros and three ones?*

Solution. The total number of distinct permutations or arrangements is

$$P_{8:5,3} = \frac{8!}{5!\,3!} = 56.$$

■

Combinations

Definition 1.2.3. *A **combination** is a set of elements without repetition and without regard to order.*

Theorem 1.2.4. *The number of combinations of n things taken k at a time is given by*

$$C_{n,k} = \frac{P_{n,k}}{k!} = \binom{n}{k} = \frac{n!}{(n-k)!\,k!}. \tag{1.43}$$

Proof. We have

$$C_{n,k} = \frac{\text{number of permutations of } n \text{ things taken } k \text{ at a time}}{\text{number of ways of reordering } k \text{ things}}.$$

■

Note that $C_{n,k} = P_{n:n-k,k}$.

Example 1.2.12. *From a rural community of 40 people, four people are selected to serve on a committee. Those selected are to serve as president, vice president, treasurer, and secretary. Find the number of committees that can be formed.*

Solution. There are $C_{40,4}$ ways of choosing an unordered group of four people from the community of forty. In addition, there are 4! ways of reordering (assigning offices) each group of four; consequently, there are

$$4! \times C_{40,4} = 4!\frac{40!}{(40-4)!\,4!} = 40 \times 39 \times 38 \times 37 = 2,193,360$$

committees.

■

One useful application of the number of combinations is the proof of the Binomial Theorem.

Theorem 1.2.5 (Binomial Theorem). *Let x and y be real numbers and let n be a positive integer. Then*

$$(x+y)^n = \sum_{k=0}^{n} \binom{n}{k} x^{n-k} y^k. \tag{1.44}$$

Proof. We have

$$(x+y)^n = \underbrace{(x+y)(x+y)\cdots(x+y)}_{n \text{ terms}},$$

a product of n sums. When the product is expanded as a sum of products, out of each term we choose either an x or a y; letting k denote the number of y's, we then have $n - k$ x's to obtain a general term of the form $x^{n-k}y^k$. There are $\binom{n}{k}$ such terms. The desired result follows by summing over $k = 0, 1, \ldots, n$. ∎

Example 1.2.13. *From four resistors and three capacitors, find the number of three component series circuits that can be formed consisting of two resistors and one capacitor. Assume each of the components is unique, and the ordering of the elements in the circuit is unimportant.*

Solution. This problem consists of a sequence of three experiments. The first and second consist of drawing a resistor, and the third consists of drawing a capacitor.

One approach is to combine the first two experiments—the number of combinations of two resistors from four is $C_{4,2} = 6$. For the third experiment, the number of combinations of one capacitor from three is $C_{3,1} = 3$. We find the number of circuits that can be found with two resistors and one capacitor to be $C_{4,2}C_{3,1} = 6 \times 3 = 18$.

For another approach, consider a typical "draw" of components to be $R_1 R_2 C$. R_1 can be any of four values, R_2 can be any of the remaining three values, and C can be any of three values—for a total number of $4 \times 3 \times 3 = 36$ possible draws of components. There are 2! ways to reorder R_1 and R_2 and 1! way of reordering the capacitor C, so that we have

$$\frac{4 \times 3 \times 3}{2!\,1!} = 18$$

possible circuits. ∎

As the previous example showed, solving a counting problem involves more than simply applying a formula. First, the problem must be clearly understood. The first step in the solution is establishing whether the problem involves a permutation or combination of the sample space. In many cases, it is convenient to utilize a tree diagram to subdivide the sample space into mutually exclusive parts, and to attack each of these parts individually. Typically, this simplifies the analysis sufficiently so that the previously developed formulas can be applied.

Combined Experiments

We now combine the concepts of permutations and combinations to provide a general framework for finding the number of samples (ordered lists) as well as the number of sets of elements (unordered) satisfying certain criteria.

Consider an experiment ε consisting of a sequence of the k subexperiments $\varepsilon_1, \varepsilon_2, \ldots, \varepsilon_k$, with outcome of the ith subexperiment denoted as $\zeta_i \in S_i$, where S_i denotes the outcome space for the ith subexperiment. We denote this combined experiment by the cartesian product:

$$\varepsilon = \varepsilon_1 \times \varepsilon_2 \times \cdots \times \varepsilon_k, \tag{1.45}$$

with outcome

$$\zeta = (\zeta_1, \zeta_2, \ldots, \zeta_k) \in S \tag{1.46}$$

and outcome space S the cartesian product

$$S = S_1 \times S_2 \times \cdots \times S_k. \tag{1.47}$$

In general,[3] the set of outcomes S_i for the ith subexperiment may depend on the outcomes $\zeta_1, \zeta_2, \ldots, \zeta_{i-1}$ which occur in the preceding subexperiments, as in sampling without replacement. The total number of possible outcomes in S is

$$n_S = n_{S_1} n_{S_2} \ldots n_{S_k}, \tag{1.48}$$

where n_{S_i} is the number of elements in S_i.

For sampling with replacement from a set of n elements, let S_1 denote the set of n elements, $S_i = S_1$, $n_{S_i} = n$, for $i = 1, 2, \ldots, k$. Consequently, there are

$$n_S = n^k \tag{1.49}$$

samples of size k with replacement from a set of n objects.

For sampling without replacement from a set of n elements, let S_1 denote the set of n elements, $S_i = S_{i-1} \cap \{\zeta_{i-1}\}^c$, for $i = 2, 3, \ldots, k$. Hence, there are

$$n_S = n \times (n - 1) \cdots (n - k + 1) = P_{n,k} \tag{1.50}$$

samples of size k without replacement from a set of n objects.

Now, consider the set $A \subset S$ with

$$A = A_1 \times A_2 \times \cdots \times A_k, \tag{1.51}$$

so that $\zeta \in A$ iff (if and only if)

$$\zeta_i \in A_i \subset S_i, \quad i = 1, 2, \ldots, k. \tag{1.52}$$

We find easily that the total number of outcomes in A is

$$n_A = n_{A_1} n_{A_2} \ldots n_{A_k}, \tag{1.53}$$

where n_{A_i} is the total number of elements in A_i, $i = 1, 2, \ldots, k$. For reasons to be made clear later, we shall refer to this set A as a **sequence event**, and often denote it simply as

$$A = A_1 A_2 \ldots A_k. \tag{1.54}$$

---

[3] Actually, S_i could depend on any or all of $\zeta_1, \zeta_2, \ldots, \zeta_k$; however, to simplify our treatment, we restrict attention to sequence experiments which admit a step-by-step implementation.

Note that n_A is the total number of outcomes in the (ordered) sequence event A.

If k_1 of the A_is are of one kind, k_2 of the A_is are of a second kind, ..., and k_r of the A_is of the rth kind, with

$$k = \sum_{i=1}^{r} k_i, \qquad (1.55)$$

then the total number of sequence events which are equivalent to A is $P_{k:k_1,k_2,...,k_r}$ so that the total number of (ordered) outcomes equivalent to the sequence event A is

$$n_{\text{tot}} = n_A \times P_{k:k_1,k_2,...,k_r} = k! \, \frac{n_{A_1} n_{A_2} \cdots n_{A_k}}{k_1! \, k_2! \, \ldots k_r!}. \qquad (1.56)$$

Finally, if the ordering of the A_is is unimportant, we find the number of distinct combinations which are equivalent to the sequence event A is

$$n_{\text{comb}} = \frac{n_{\text{tot}}}{k!} = \frac{n_{A_1} n_{A_2} \cdots n_{A_k}}{k_1! \, k_2! \, \ldots k_r!}, \qquad (1.57)$$

where $k!$ is the number of ways of reordering a k-dimensional outcome.

Example 1.2.14.   *Five cards are dealt from a standard 52 card deck of playing cards. There are four suits (Hearts, Spades, Diamonds, and Clubs), with 13 cards in each suit (2,3,4,5,6,7,8,9,10, Jack, Queen, King, Ace). (a) How many five-card hands can be dealt? (b) How many five-card hands contain exactly two Hearts? (c) How many hands contain exactly one Jack, two Queens, and two Aces?*

Solution

(a) There are $P_{52,5} = 52 \times 51 \times 50 \times 49 \times 48 \approx 3.12 \times 10^8$ five-card hands, counting different orderings as distinct. Since the ordering of cards in the hand is not important, there are

$$C_{52,5} = \frac{P_{52,5}}{5!} = 2598960 \approx 2.6 \times 10^6$$

five-card hands that can be drawn.

(b) Consider the sequence event

$$A = H_1 H_2 X_3 X_4 X_5,$$

where H_i denotes a heart on the ith draw and X_i denotes a non-heart drawn on the ith draw. There are

$$n_A = 13 \times 12 \times 39 \times 38 \times 37 = 8554104$$

outcomes in the sequence event A. Of the five cards, two are of type heart, and three are of type non-heart, for a total of

$$n_{tot} = n_A P_{5:2,3} = n_A \times 10 = 85541040$$

outcomes equivalent to those in A. Finally, since the ordering of cards in a hand is unimportant, we find that there are

$$\frac{n_A}{2!\,3!} = \frac{13 \times 12 \times 39 \times 38 \times 37}{2!\,3!} = 712842$$

hands with exactly two hearts.

An alternative is to compute

$$C_{13,2}C_{39,3} = \frac{13 \times 12}{2!}\frac{39 \times 38 \times 37}{3!} = 712842.$$

(c) Consider the sequence event

$$B = J_1 Q_2 Q_3 A_4 A_5.$$

Arguing as in (b), we find that there are

$$\frac{4 \times 4 \times 3 \times 4 \times 3}{1!\,2!\,2!} = 144$$

hands with one Jack, two Queens, and two Aces.

An alternative is to compute

$$C_{4,1}C_{4,2}C_{4,2} = \frac{4}{1}\frac{4 \times 3}{2}\frac{4 \times 3}{2} = 144.$$ ∎

Example 1.2.15. *Suppose a committee of four members is to be formed from four boxers (A,B,C,D), five referees (E,F,G,H,I) and TV announcer J. Furthermore, A and B hate each other and cannot be on the same committee unless it contains a referee. How many committees can be formed?*

Solution. This problem can best be solved by reducing the problem into a mutually exclusive listing of smaller problems. There are four acceptable committee compositions:

1. A is on the committee and B is not. Consider the sequence event $AX_2X_3X_4$, where X_i consists of all previously unselected candidates except B. There are $1 \times 8 \times 7 \times 6$ outcomes in the sequence event $AX_2X_3X_4$. Since X_2, X_3, and X_4, are equivalent and the order is unimportant, we find that there are

$$\frac{1 \times 8 \times 7 \times 6}{1!\,3!} = 56$$

distinct committees with A on the committee and B not on the committee.

2. B is on the committee and A is not. This committee composition is treated as case 1 above with A and B interchanged. There are thus 56 such committees.

3. Neither A nor B is on the committee. Consider the sequence event $X_1 X_2 X_3 X_4$, where X_i denotes any previously unselected candidate except A or B. There are

$$\frac{8 \times 7 \times 6 \times 5}{0! \, 4!} = 70$$

such committees.

4. A and B are on the committee, along with at least one referee. Consider $ABRX_4$, where $X_4 = \{C, D, J\}$. There are

$$\frac{1 \times 1 \times 5 \times 3}{1! \, 1! \, 1! \, 1!} = 15$$

such committees with one referee. By considering ABR_3R_4, there are

$$\frac{1 \times 1 \times 5 \times 4}{1! \, 1! \, 2!} = 10$$

distinct committees with two referees.

Thus, the total number of acceptable committees is

$$56 + 56 + 70 + 15 + 10 = 207. \qquad \blacksquare$$

Example 1.2.16. *A change purse contains five nickels, eight dimes, and three quarters. Find the number of ways of drawing two quarters, three dimes, and four nickels if:*

(a) *coins are distinct and order is important,*

(b) *coins are distinct and order is unimportant,*

(c) *coins are not distinct and order is important,*

(d) *coins are not distinct and order is unimportant.*

Solution. Let sequence event

$$A = Q_1 Q_2 D_3 D_4 D_5 N_6 N_7 N_8 N_9.$$

Sequence event A represents one way to obtain the required collection of coins. There are a total of $3 \times 2 \times 8 \times 7 \times 6 \times 5 \times 4 \times 3 \times 2 = 241920$ outcomes in A. There are

$$\frac{9!}{2! \, 3! \, 4!} = \frac{9 \times 8 \times 7 \times 6 \times 5}{2 \times 3 \times 2} = 2 \times 630 = 1260$$

ways to reorder the different *types* of coins in A, so that there are a total of 1260 sequence events which, like A, contain two quarters, three dimes, and four nickels.

(a) We find that there are

$$241920 \times \frac{9!}{2!\,3!\,4!} \approx 3.048 \times 10^8$$

ways to draw two quarters, three dimes, and four nickels if the coins are distinct and order is important.

(b) There are 9! ways to reorder nine distinct items, so that there are

$$\frac{241920}{9!} \times \frac{9!}{2!\,3!\,4!} = \frac{241920}{2!\,3!\,4!} = 840$$

ways to draw two quarters, three dimes, and four nickels if the coins are distinct and order is unimportant.

(c) Sequence event A represents one way to obtain the required collection of coins. There are

$$\frac{9!}{2!\,3!\,4!} = 1260$$

ways to reorder the different *types* of coins in A, so that there are 1260 ways to draw two quarters, three dimes, and four nickels if the coins are not distinct and order is important.

(d) There is 1 way to draw two quarters, three dimes, and four nickels if the coins are not distinct and order is unimportant.   ∎

Drill Problem 1.2.1. *An urn contains four balls labeled 0, 1, 2, 3. Two balls are selected one after the other without replacement. Enumerate the sample space using both a tree diagram and a coordinate system.*

Answers: 12, 12.

Drill Problem 1.2.2. *An urn contains three balls labeled 0, 1, 2. Reach in and draw one ball to determine how many times a coin is to be flipped. Enumerate the sample space with a tree diagram.*

Answer: 7.

Drill Problem 1.2.3. *Professor S. Rensselaer teaches a course in probability theory. She is a kind-hearted but very tricky old lady who likes to give many unannounced quizzes during the week. She determines the number of quizzes each week by tossing a fair tetrahedral die with faces labeled 1, 2, 3, 4. The more quizzes she gives, however, the less time she has to assign and grade homework problems. If Ms. Rensselaer is to give L quizzes during the week, then she will assign from 1 to 5-L homework problems. Enumerate the sample space describing the number of quizzes and homework problems she gives each week.*

Answer: 10.

Drill Problem 1.2.4. *Determine the number of 8-bit computer words that can be formed if: (a) the first character is zero, (b) the last two characters are one, (c) the first character is zero and the last two characters are one, (d) all of the characters are zero.*

Answers: 1, 32, 64, 128.

Drill Problem 1.2.5. *Determine the number of even three digit numbers that can be formed from S if each digit can be used only once and (a) $S = \{1, 2, 4, 7\}$; (b) $S = \{2, 4, 6, 8\}$; (c) $S = \{1, 2, 3, 5, 7, 8, 9\}$; (d) $S = \{1, 3, 5, 7, 9\}$.*

Answers: 24, 12, 60, 0.

Drill Problem 1.2.6. *Eight students A through H enter a student paper contest in which awards are given for first, second and third place. Determine the number of finishes (a) possible; (b) if student C is awarded first place; (c) if students C and D are given an award; (d) if students C or D are given an award, but not both.*

Answers: 336, 42, 36, 180.

Drill Problem 1.2.7. *Determine the number of 8-bit computer words containing (a) three zeros; (b) three zeros, given the first bit is one; (c) two zeros, given the first three bits are one; (d) more zeros than ones.*

Answers: 35, 56, 93, 10.

Drill Problem 1.2.8. *Suppose a committee consisting of three members is to be formed from five men and three women. How many committees (a) can be formed; (b) can be formed with two women and one man; (c) can be formed with one woman and two men if a certain man must be on the committee?*

Answers: 15, 56, 12.

Drill Problem 1.2.9. *A college plays eight conference and two nonconference football games during a season. Determine the number of ways the team may end the season with (a) eight wins and two losses; (b) three wins, six losses, and one tie; (c) at least seven wins and no ties; (d) three wins in their first four games and two wins and three losses in the remaining games.*

Answers: 840, 176, 480, 45.

1.3   DEFINITION OF PROBABILITY

Up to this point, we have discussed an experiment and the outcome space for the experiment. We have devoted some effort to evaluating the number of specific types of outcomes in the case of a discrete outcome space. To a large extent, probability theory provides analytical methods for assigning and/or computing the "likelihood" that various "phenomena" associated with the experiment occur. Since the experimental outcome space is the set of all possible outcomes of the experiment, it is clear that any phenomenon for which we have an interest may be considered to be some subset of the outcome space. We will henceforth refer to such a subset of S as an **event**. We say that event A has occurred if the experimental outcome $\zeta \in A$. Thus, if $A \subset S$ is an event we denote the probability (or likelihood) that event A has occurred as $P(A)$. While *any* subset of S is a potential event—we will find that a large simplification occurs when we investigate only those events in which we have some interest. Let's agree at the outset that $P(A)$ is a real number between 0 and 1, with $P(A) = 0$ meaning that the event A is extremely unlikely to occur and $P(A) = 1$ meaning that the event A is almost certain to occur.

Several approaches to probability theory have been taken. Four approaches will be discussed here: classical, relative frequency, personal probability and axiomatic.

1.3.1   Classical

The classical approach to probability evolved from the gambling dens of Europe in the 1600s. It is based on the idea that any experiment can be broken down into a fine enough space so that each single outcome is equally likely. All events are then made up of the mutually exclusive outcomes. Thus, if the total number of outcomes is N and the event A occurs for N_A of these outcomes, the classical approach defines the probability of A,

$$P(A) = \frac{N_A}{N}. \tag{1.58}$$

This definition suffers from an obvious fault of being circular. The statement of being "equally likely" is actually an assumption of certain probabilities. Despite this and other faults, the classical definition works well for a certain class of problems that come from games of chance or are similar in nature to games of chance. We will use the classical definition in assuming certain probabilities in many of our examples and problems, but we will not develop a theory of probability from it.

1.3.2   Relative Frequency

The relative frequency definition of probability is based on observation or experimental evidence and not on prior knowledge. If an experiment is repeated N times and a certain event A occurs in N_A of the trials, then the probability of A is defined to be

$$P(A) = \lim_{N \to \infty} \frac{N_A}{N}. \tag{1.59}$$

For example, if a six-sided die is rolled a large number of times and the numbers on the face of the die come up in approximately equal proportions, then we could say that the probability of each number on the upturned face of the die is 1/6. The difficulty with this definition is determining when N is sufficiently large and indeed if the limit actually exists. We will certainly use this definition in relating deduced probabilities to the physical world, but we will not develop probability theory from it.

1.3.3   Personal Probability

Personal or subjective probability is often used as a measure of belief whether or not an event may have occurred or is going to occur. Its only use in probability theory is to subjectively determine certain probabilities or to subjectively interpret resulting probability calculations. It has no place in the development of probability theory.

1.3.4   Axiomatic

Last, we turn our attention to the axiomatic[4] definition of probability in which we assign a number, called a probability, to each event in the event space. For now, we consider the event space (denoted by $\mathcal{F}$) to be simply the space containing all events to which we wish to assign a probability. Logically, the probability that event A occurs should relate to some physical average not conflicting with the other definitions of probability, and should reflect the chance of that event occurring in the performance of the experiment. Given this assignment, the axiomatic definition of probability is stated as follows. We assign a probability to each event in the event space according to the following axioms:

A1: $P(A) \geq 0$ for any event $A \in \mathcal{F}$;

A2: $P(S) = 1$;

A3: If $A_1, A_2, \ldots$ are mutually exclusive events in $\mathcal{F}$, then

$$P\left(\bigcup_{i=1}^{\infty} A_i\right) = \sum_{i=1}^{\infty} P(A_i).$$

When combined with the properties of the event space $\mathcal{F}$ (treated in the following section), these axioms are all that is necessary to derive all of the theorems of probability theory.

Consider the third axiom. Let $A_1 = S$ and $A_i = \varnothing$, for $i > 1$. Then $A_1, A_2, \ldots$ are mutually exclusive and the third axiom yields

$$P(S) = P(S) + \sum_{i=2}^{\infty} P(\varnothing)$$

so that $P(\varnothing) = 0$. Now, with $A_1 = A$, $A_2 = B$, $A_i = \varnothing$ for $i > 2$, and $A \cap B = \varnothing$, the third axiom yields $P(A \cup B) = P(A) + P(B)$.

---

[4]An **axiom** is a self-evident truth or proposition; an established or universally received principle.

Some additional discussion concerning the selection of the three axioms is in order. Both the classical definition and the relative frequency definition give some physical meaning to probability. The axiomatic definition agrees with this. Consider the axioms one at a time. Certainly, the first axiom does not conflict since probability is nonnegative for the first two definitions. The second axiom also agrees since if event A always occurs, then $N_A = N$ and $P(A) = 1$. But why the third axiom? Consider the classical definition with two events A and B occurring in N_A and N_B outcomes, respectively. With A and B mutually exclusive, the total number of events in which A or B occurs is $N_A + N_B$. Therefore,

$$P(A \cup B) = \frac{N_A + N_B}{N} = P(A) + P(B),$$

which agrees with the third axiom. If A and B are not mutually exclusive, then both A and B could occur for some outcomes and the total number of outcomes in which A or B occurs is less than $N_A + N_B$. A similar argument can be made with the relative frequency definition.

The axioms do not tell us how to compute the probabilities for each sample point in the sample space—nor do the axioms tell us how to assign probabilities for each event in the event space $\mathcal{F}$. Rather, the axioms provide a consistent set of rules which must be obeyed when making probability assignments. Either the classical or relative frequency approach is often used for making probability assignments.

One method of obtaining a probability for each sample point in a finite sample space is to assign a weight, w, to each sample point so that the sum of all weights is one. If the chance of a particular sample point occurring during the performance of the experiment is quite likely, the weight should be close to one. Similarly, if the chance of a sample point occurring is unlikely, the weight should be close to zero. When this chance of occurrence is determined by experimentation, we are using the relative frequency definition. If the experiment has a sample space in which each outcome is equally likely, then each outcome is assigned an equal weight according to the classical definition. After we have assigned a probability to each of the outcomes, we can find the probability of any event by summing the probabilities of all outcomes included in the event. This is a result of axiom three, since the outcomes are mutually exclusive, single element events.

Example 1.3.1. *A die is tossed once. What is the probability of an even number occurring?*

Solution. The sample space for this experiment is

$$S = \{1, 2, 3, 4, 5, 6\}.$$

Since the die is assumed fair, each of these outcomes is equally likely to occur. Therefore, we assign a weight of w to each sample point; i.e., $P(i) = w$, $i = 1, 2, \ldots, 6$. By the second and third axioms of probability we have $P(S) = 1 = 6w$; hence, $w = 1/6$. Letting $A = \{2, 4, 6\}$, we find the probability of event A equals

$$P(A) = P(\{2\}) + P(\{4\}) + P(\{6\}) = \frac{1}{2}. \qquad \blacksquare$$

Example 1.3.2.   *A tetrahedral die (with faces labeled 0,1,2,3) is loaded so that the zero is three times as likely to occur as any other number. If A denotes the event that an odd number occurs, then find $P(A)$ for one toss of the die.*

Solution.   The sample space for this experiment is $S = \{0, 1, 2, 3\}$. Assigning a weight w to the sample points 1, 2, and 3; and $3w$ to zero, we find

$$P(S) = 1 = P(\{0\}) + P(\{1\}) + P(\{2\}) + P(\{3\}) = 3w + w + w + w = 6w,$$

and $w = 1/6$. Thus $P(A) = P(\{1\}) + P(\{3\}) = 1/3$. $\qquad \blacksquare$

Example 1.3.3.   *Find the probability of exactly four zeros occurring in an 8-bit word.*

Solution.   From the previous section, we know that there are $2^8 = 256$ outcomes in the sample space. Let event $A = \{00001111\}$. Since each outcome is assumed equally likely, we have $P(A) = 1/256$. We need to multiply $P(A)$ by the number of events which (like A) have exactly four zeros and four ones; i.e., by

$$P_{8:4,4} = \frac{8!}{4!4!} = 70.$$

So, the desired probability is $70/256$. $\qquad \blacksquare$

Example 1.3.4.   *A fair coin is tossed twice. If A is the event that at least one head appears, and B is the event that two heads appear, find $P(A \cup B)$.*

Solution.   Letting H_i and T_i denote a Head and Tail, respectively, on the ith toss, we find that

$$A = \{H_1 H_2, H_1 T_2, T_1 H_2\},$$
$$B = \{H_1 H_2\} \subset A;$$

hence, $P(A \cup B) = P(A) = 3/4$. It is important to note that in this case $P(A) + P(B) = 1 \neq P(A \cup B)$ since the events A and B are not mutually exclusive. $\qquad \blacksquare$

Example 1.3.5.   *Three cards are drawn at random (each possibility is equally likely) from an ordinary deck of 52 cards (without replacement). Find the probability p that two are spades and one is a heart.*

Solution.   There are a total of $52 \times 51 \times 50$ possible outcomes of this experiment. Consider the sequence event $A = S_1 S_2 H_3$, denoting a spade drawn on each of the first two draws, and a heart on the third draw. There are $13 \times 12 \times 13$ outcomes in the sequence event A. There are

$$\frac{3!}{2! \, 1!} = 3$$

mutually exclusive events which, like A, contain two spades and one heart. We conclude that

$$p = \frac{13 \times 12 \times 13 \times 3}{52 \times 51 \times 50} = \frac{39}{850}.$$

An alternative is to compute

$$p = \frac{C_{13,2}C_{13,1}}{C_{52,3}} = \frac{39}{850}. \qquad \blacksquare$$

The preceding examples illustrate a very powerful technique for computing probabilities: express the desired event as a union of mutually exclusive events with known probabilities and apply the third axiom of probability. As long as all such events are in the event space $\mathcal{F}$, this technique works well. When the outcome space is discrete, the event space can be taken to be the collection of *all* subsets of S; however, when the outcome space S contains an uncountably infinite number of elements the technique fails. For now, we assume all needed events are in the event space and address the necessary structure of the event space in the next section. The following theorem, which is a direct consequence of the axioms of probability, provides additional analytical ammunition for attacking probability problems.

Theorem 1.3.1. *Assuming that all events indicated are in the event space $\mathcal{F}$, we have:*

(i) $P(A^c) = 1 - P(A)$,
(ii) $P(\varnothing) = 0$,
(iii) $0 \leq P(A) \leq 1$,
(iv) $P(A \cup B) = P(A) + P(B) - P(A \cap B)$, *and*
(v) $P(B) \leq P(A)$ *if* $B \subset A$.

Proof

(i) Since $S = A \cup A^c$ and $A \cap A^c = \varnothing$, we apply the second and third axioms of probability to obtain

$$P(S) = 1 = P(A) + P(A^c),$$

from which (i) follows.
(ii) Applying (i) with $A = S$ we have $A^c = \varnothing$ so that $P(\varnothing) = 1 - P(S) = 0$.
(iii) From (i) we have $P(A) = 1 - P(A^c)$, from the first axiom we have $P(A) \geq 0$ and $P(A^c) \geq 0$; consequently, $0 \leq P(A) \leq 1$.

(iv) Let $C = B \cap A^c$. Then

$$A \cup C = A \cup (B \cap A^c) = (A \cup B) \cap (A \cup A^c) = A \cup B,$$

and $A \cap C = A \cap B \cap A^c = \varnothing$, so that $P(A \cup B) = P(A) + P(C)$. Furthermore, $B = (B \cap A) \cup (B \cap A^c)$ and $(B \cap A) \cap (B \cap A^c) = \varnothing$ so that $P(B) = P(C) + P(A \cap B)$ and $P(C) = P(B) - P(A \cap B)$.

(v) Since $B \subset A$, we have $A = (A \cap B) \cup (A \cap B^c) = B \cup (A \cap B^c)$. Consequently,

$$P(A) = P(B) + P(A \cap B^c) \geq P(B). \qquad \blacksquare$$

The above theorem and its proof are extremely important. The reader is urged to digest it totally—Venn diagrams are permitted to aid in understanding.

Example 1.3.6.   *Given $P(A) = 0.4$, $P(A \cap B^c) = 0.2$, and $P(A \cup B) = 0.6$, find $P(A \cap B)$ and $P(B)$.*

Solution. We have $P(A) = P(A \cap B) + P(A \cap B^c)$ so that $P(A \cap B) = 0.4 - 0.2 = 0.2$. Similarly,

$$P(B^c) = P(B^c \cap A) + P(B^c \cap A^c) = 0.2 + 1 - P(A \cup B) = 0.6.$$

Hence, $P(B) = 1 - P(B^c) = 0.4$. $\qquad \blacksquare$

Example 1.3.7.   *A man is dealt four spade cards from an ordinary deck of 52 playing cards, and then dealt three additional cards. Find the probability p that at least one of the additional cards is also a spade.*

Solution. We may start the solution with a 48 card deck of 9 spades and 39 non-spade cards.

One approach is to consider all sequence events with at least one spade: $S_1 N_2 N_3$, $S_1 S_2 N_3$, and $S_1 S_2 S_3$, along with the reorderings of these events.

Instead, consider the sequence event with no spades: $N_1 N_2 N_3$, which contains $39 \times 38 \times 37$ outcomes. We thus find

$$1 - p = \frac{39 \times 38 \times 37}{48 \times 47 \times 46} = \frac{9139}{17296},$$

or $p = 8157/17296$. $\qquad \blacksquare$

Boole's Inequality below provides an extension of Theorem 1.3.1(iv) to the case with many non-mutually exclusive events.

Theorem 1.3.2 (Boole's Inequality). *Let $A_1, A_2, \ldots$ all belong to $\mathcal{F}$. Then*

$$P\left(\bigcup_{i=1}^{\infty} A_i\right) = \sum_{k=1}^{\infty}(P(A_k) - P(A_k \cap B_k)) \le \sum_{k=1}^{\infty} P(A_k),$$

where

$$B_k = \bigcup_{i=1}^{k-1} A_i.$$

Proof.   Note that $B_1 = \varnothing$, $B_2 = A_1$, $B_3 = A_1 \cup A_2, \ldots$, $B_k = A_1 \cup A_2 \cup \cdots \cup A_{k-1}$; as k increases, the size of B_k is nondecreasing. Let $C_k = A_k \cap B_k^c$; thus,

$$C_k = A_k \cap \left(A_1^c \cap A_2^c \cap \cdots \cap A_{k-1}^c\right)$$

consists of all elements in A_k and not in any A_i, $i = 1, 2, \ldots, k - 1$. Then

$$B_{k+1} = \bigcup_{i=1}^{k} A_i = B_k \cup \underbrace{(A_k \cap B_k^c)}_{C_k},$$

and

$$P(B_{k+1}) = P(B_k) + P(C_k).$$

We have $P(B_2) = P(C_1)$, $P(B_3) = P(C_1) + P(C_2)$, and

$$P(B_{k+1}) = P\left(\bigcup_{i=1}^{k} A_i\right) = \sum_{i=1}^{k} P(C_i).$$

The desired result follows by noting that

$$P(C_i) = P(A_i) - P(A_i \cap B_i). \qquad \blacksquare$$

While the above theorem is useful in its own right, the proof illustrates several important techniques. The third axiom of probability requires a sequence of mutually exclusive events. The above proof shows one method for obtaining a collection of n mutually exclusive events from a collection of n arbitrary events. It often happens that one is willing to settle for an upper bound on a needed probability. The above proof may help convince the reader that such a bound might be much easier to obtain than carrying out a complete, exact analysis. It is up to the user, of course, to determine when a bound is acceptable. Obviously, when an upper bound on a probability exceeds one the upper bound reveals absolutely no relevant information!

Example 1.3.8. *Let $S = [0, 1]$ (the set of real numbers $\{x : 0 \le x \le 1\}$). Let $A_1 = [0, 0.5]$, $A_2 = (0.45, 0.7)$, $A_3 = [0.6, 0.8)$, and assume $P(\zeta \in I) =$ length of the interval $I \cap S$, so that $P(A_1) = 0.5$, $P(A_2) = 0.25$, and $P(A_3) = 0.2$. Find $P(A_1 \cup A_2 \cup A_3)$.*

Solution. Let $C_1 = A_1$, $C_2 = A_2 \cap A_1^c = (0.5, 0.7)$, and $C_3 = A_3 \cap A_1^c \cap A_2^c = [0.7, 0.8)$. Then C_1, C_2, and C_3 are mutually exclusive and $A_1 \cup A_2 \cup A_3 = C_1 \cup C_2 \cup C_3$; hence

$$P(A_1 \cup A_2 \cup A_3) = P(C_1 \cup C_2 \cup C_3) = 0.5 + 0.2 + 0.1 = 0.8.$$

Note that for this example, Boole's inequality yields

$$P(A_1 \cup A_2 \cup A_3) \le 0.5 + 0.25 + 0.2 = 0.95.$$

This is an example of an uncountable outcome space. It turns out that for this example, it is impossible to compute the probabilities for *every* possible subset of S. This dilemma is addressed in the following section. ∎

Drill Problem 1.3.1. *Let $P(A) = 0.35$, $P(B) = 0.5$, $P(A \cap B) = 0.2$, and let C be an arbitrary event. Determine: (a) $P(A \cup B)$; (b) $P(B \cap A^c)$; (c) $P((A \cap B) \cup (A \cap B^c) \cup (A^c \cap B))$; (d) $P((A \cap B^c) \cup (A^c \cap B) \cup (A^c \cap B \cap C^c))$.*

Answers: 0.45, 0.3, 0.65, 0.65.

Drill Problem 1.3.2. *A pentahedral die (with faces labeled 1,2,3,4,5) is loaded so that an even number is twice as likely to occur as an odd number (e.g., $P(\{2\}) = 2P(\{1\})$). Let A equal the event that a number less than three occurs and B equal the event that the number is even. Determine: (a) $P(A)$; (b) $P(B)$; (c) $P(A^c \cup B^c)$; (d) $P(A \cup (B \cap A^c))$.*

Answers: 3/7, 4/7, 5/7, 5/7.

Drill Problem 1.3.3. *A woman is dealt two hearts and a spade from a deck of cards. She is given four more cards. Determine the probability that: (a) one is a spade; (b) two are hearts; (c) two are spades and one is a club; (d) at least one is a club.*

Answers: 0.18249, 0.09719, 0.72198, 0.44007.

Drill Problem 1.3.4. *Determine the probability that an 8-bit computer word contains: (a) four zeros; (b) four zeros, given the last bit is a 1; (c) two zeros, given the first three bits are one; (d) more zeros than ones.*

Answers: 70/256, 70/256, 80/256, 93/256.

1.4   THE EVENT SPACE

Although the techniques presented in the previous section are always possible when the experimental outcome space is discrete, they fall short when the outcome space is not discrete. For example, consider an experiment with outcome any real number between 0 and 5, with all numbers equally likely. We then have that the probability of any specific number between 0 and 5 occurring is exactly 0. Attempting to let the event space be the collection of all subsets of the outcome space $S = \{x : 0 \leq x \leq 5\}$ then leads to serious difficulties in that it is impossible to assign a probability to each event in this event space. By reducing our ambitions with the event space, we will see in this section that we will be able to come up with an event space which is rich enough to enable the computation of the probability for any event of practical interest.

Definition 1.4.1. *A collection $\mathcal{F}$ of subsets of S is a **field** (or algebra) of subsets of S if the following properties are all satisfied:*

F1: $\varnothing \in \mathcal{F}$,

F2: If $A \in \mathcal{F}$ then $A^c \in \mathcal{F}$, and

F3: If $A_1 \in \mathcal{F}$ and $A_2 \in \mathcal{F}$ then $A_1 \cup A_2 \in \mathcal{F}$.

Example 1.4.1. *Consider a single die toss experiment. (a) How many possible events are there? (b) Is the collection of all possible subsets of S a field? (c) Consider $\mathcal{F} = \{\varnothing, \{1, 2, 3, 4, 5, 6\}, \{1, 3, 5\}, \{2, 4, 6\}\}$. Is $\mathcal{F}$ a field?*

Solution. (a) Using the Binomial Theorem, we find that there are

$$n = \sum_{k=0}^{6} C_{6,k} = (1 + 1)^6 = 64$$

possible subsets of S. Hence, the number of possible events is 64. Note that there are only six possible outcomes.

Each of the collections (b) and (c) is a field, as can readily be seen by checking F1, F2, and F3 above. ∎

Theorem 1.4.1. *Let $A_1, A_2, \ldots, A_n$ all belong to the field $\mathcal{F}$. Then*

$$\bigcup_{i=1}^{n} A_i \in \mathcal{F}$$

and

$$\bigcap_{i=1}^{n} A_i \in \mathcal{F}$$

Proof. Let

$$B_k = \bigcup_{i=1}^{k} A_i, \quad k = 1, 2, \ldots, n.$$

Then from F3 we have $B_2 \in \mathcal{F}$. But then $B_3 = A_3 \cup B_2 \in \mathcal{F}$. Assume $B_{k-1} \in \mathcal{F}$ for some $2 \leq k < n$. Then using F3 we have $B_k = A_k \cup B_{k-1} \in \mathcal{F}$; hence, $B_k \in \mathcal{F}$ for $k = 1, 2, \ldots, n$. Using F2, $A_1^c, A_2^c, \ldots, A_n^c$ are all in $\mathcal{F}$. The above then shows that

$$\bigcup_{i=1}^{k} A_i^c \in \mathcal{F}, \quad k = 1, 2, \ldots, n.$$

Finally, using F2 and De Morgan's Law:

$$\left(\bigcup_{i=1}^{k} A_i^c \right)^c = \bigcap_{i=1}^{k} A_i \in \mathcal{F}. \quad k = 1, 2, \ldots, n. \qquad \blacksquare$$

The above theorem guarantees that finite unions and intersections of members of a field $\mathcal{F}$ also belong to $\mathcal{F}$. The following example demonstrates that countably infinite unions of members of a field $\mathcal{F}$ are not necessarily in $\mathcal{F}$.

Example 1.4.2. *Let $\mathcal{F}$ be the field of subsets of real numbers containing sets of the form*

$$G_a = (-\infty, a].$$

(a) Find G_a^c. (b) Find $G_a \cup G_b$. (c) Find $G_a^c \cap G_b$. (d) Simplify

$$A = \bigcup_{n=1}^{\infty} \left(-\infty, a - \frac{1}{n} \right].$$

Is $A \in \mathcal{F}$?

Solution

(a) With $S = R$ (the set of all real numbers), we have

$$G_a^c = \{x \in S : x \notin G_a\} = \{x : a < x < \infty\} = (a, \infty).$$

(b) Using the definition of set union,

$$G_a \cup G_b = \{x \in R : x \leq a \text{ or } x \leq b\} = (-\infty, \max\{a, b\}].$$

(c) Using the definition of set intersection,

$$G_a^c \cap G_b = (a, \infty) \cap (-\infty, b] = (a, b].$$

(d) We find

$$A = (-\infty, a - 1] \cup (-\infty, a - 1/2] \cup (-\infty, a - 1/3] \cdots$$

so that $A = (-\infty, a) \notin \mathcal{F}$. $\qquad \blacksquare$

Definition 1.4.2. *A collection $\mathcal{F}$ of subsets of S is a **sigma-field** (or **sigma-algebra**) of subsets of S if F1, F2, and F3a are all satisfied, where*
    F3a: *If $A_1, A_2, \ldots$ are all in $\mathcal{F}$, then*

$$\bigcup_{i=1}^{\infty} A_i \in \mathcal{F}.$$

Theorem 1.4.2. *Let $\mathcal{F}$ be a σ-field of subsets of S. Then*

(i) *$\mathcal{F}$ is a field of subsets of S, and*
(ii) *If $A_i \in \mathcal{F}$ for $i = 1, 2, \ldots$, then*

$$\bigcap_{i=1}^{\infty} A_i \in \mathcal{F}.$$

Proof. (i) Let $A_1 \in \mathcal{F}$, $A_2 \in \mathcal{F}$, and $A_n = \varnothing$ for $n = 3, 4, \ldots$. Then F1 and F3a imply that

$$\bigcup_{i=1}^{\infty} A_i = A_1 \cup A_2 \in \mathcal{F},$$

so that F3 is satisfied; hence, $\mathcal{F}$ is a field. The proof of (ii) is similar to the previous theorem, following from De Morgan's Law. ∎

Definition 1.4.3. *Let A be any collection of subsets of S. We say that a σ-field $\mathcal{F}_0$ of subsets of S is a **minimal σ-field over** A (denoted $\sigma(A)$) if $A \subset \mathcal{F}_0$ and if $\mathcal{F}_0$ is contained in every σ-field that contains A.*

Theorem 1.4.3. *$\sigma(A)$ exists for any collection A of subsets of S.*

Proof. Let C be the collection of all σ-fields of subsets of S that contain A. Since the collection of all subsets of S is a σ-field containing A, C is nonempty. Let

$$\mathcal{F}_0 = \bigcap_{\mathcal{F} \in C} \mathcal{F}.$$

Since $\varnothing \in \mathcal{F}$ for all $\mathcal{F} \in C$, we have $\varnothing \in \mathcal{F}_0$. If $A \in \mathcal{F}_0$, then $A \in \mathcal{F}$ for all $\mathcal{F} \in C$ so that $A^c \in \mathcal{F}$ for all $\mathcal{F} \in C$; hence $A^c \in \mathcal{F}_0$. If $A_1, A_2, \ldots$ are all in $\mathcal{F}_0$ then $A_1, A_2, \ldots$ are all in $\mathcal{F}$ for every $\mathcal{F} \in C$ so that

$$\bigcup_{i=1}^{\infty} A_i \in \mathcal{F}_0.$$

Consequently, $\mathcal{F}_0$ is a σ-field of subsets of S that contains A. We conclude that $\mathcal{F}_0 = \sigma(A)$, the minimal σ-field of subsets of S that contains A.   ∎

As the astute reader will have surmised by now, we will insist that the event space $\mathcal{F}$ be a σ-field of subsets of the outcome space S. We can tailor a special event space for a given problem by starting with a collection of events in which we have some interest. The minimal σ-field generated by this collection is then a legitimate event space, and is guaranteed to exist thanks to the above theorem. We are (fortunately) not usually required to actually *find* the minimal σ-field. Any of the standard set operations on events in this event space yield events which are also in this event space; thus, the event space is closed under the set operations of complementation, union, and intersection.

Example 1.4.3. *Consider the die-toss experiment and suppose we are interested only in the event* $A = \{1, 3, 5, 6\}$. *Find the minimal σ-field $\sigma(A)$, where $A = \{A\}$.*

Solution. We find easily that

$$\sigma(A) = \{\varnothing, S, A, A^c\}. \qquad ∎$$

A very special σ-field will be quite important in our future work with probability theory. The Borel field contains *all* sets of real numbers in which one might have a "practical interest."

Definition 1.4.4. *Let $S = \Re$ (the set of all real numbers). The minimal σ-field over the collection of open sets of $\Re$ is called a **Borel field**. Members of this σ-field are called **Borel sets**.*

It is very important to note that all sets of real numbers having practical significance are Borel sets. We use standard interval notation to illustrate. We have that

$$(a, b] = \{x \in \Re : a < x \le b\} = \bigcap_{n=1}^{\infty} \left(a, b + \frac{1}{n} \right),$$

$$[a, b) = \{x \in \Re : a \le x < b\} = \bigcap_{n=1}^{\infty} \left(a - \frac{1}{n}, b \right),$$

and

$$[a, b] = \{x \in \Re : a \le x \le b\} = \bigcap_{n=1}^{\infty} \left(a - \frac{1}{n}, b + \frac{1}{n} \right)$$

are all Borel sets; hence any countable union or intersection or complement of such sets is also a Borel set. For example, the set of all positive integers is also a Borel set. Indeed, examples of sets which are not Borel sets do not occur frequently in applications of probability. We note that the set of all irrational real numbers is not a Borel set.

Drill Problem 1.4.1. *Find the minimal sigma-field containing the events A and B.*

Answer: $\sigma(\{A, B\}) = \{\varnothing, S, A, A^c, B, B^c, A \cup B, A \cup B^c, A^c \cup B, A^c \cup B^c, A^c \cap B^c,$
$A^c \cap B, A \cap B^c, A \cap B, (A^c \cap B) \cup (A \cap B^c), (A^c \cap B^c) \cup (A \cap B)\}.$

Drill Problem 1.4.2. *Simplify:*

(a) $\displaystyle\bigcup_{n=1}^{\infty} \left(a - \frac{1}{n}, b + \frac{2}{n} \right),$

(b) $\displaystyle\bigcap_{n=1}^{\infty} \left(a - \frac{1}{n}, b + \frac{2}{n} \right).$

Answer: $(a - 1, b + 2), [a, b]$.

1.5   THE PROBABILITY SPACE

In this section, we present a few definitions from a branch of mathematics known as measure theory. These definitions along with the previous sections enable us to define a probability space. Measure theory deals with the determination of how "big" a set is—much as a ruler can measure length. A probability measure reveals how much "probability" an event has.

Definition 1.5.1. *A (real-valued) **set function** is simply a function which has a set as the independent variable; i.e., a set function is a mapping from a collection of sets to real numbers.*

Definition 1.5.2. *A set function G defined on a σ-field $\mathcal{F}$ is σ-**additive** if*

(i) $G(\varnothing) = 0$, and

(ii) If $A_1, A_2, \ldots$ are mutually exclusive members of $\mathcal{F}$ then

$$G\left(\bigcup_{n=1}^{\infty} A_n \right) = \sum_{n=1}^{\infty} G(A_n).$$

Definition 1.5.3. *Let G be a set function defined on a σ-field $\mathcal{F}$. The set function G is σ-finite if $G(S) < \infty$, and nonnegative if $G(A) \geq 0$ for all $A \in \mathcal{F}$. A nonnegative σ-additive set function G defined on a σ-field $\mathcal{F}$ is called a measure.*

Definition 1.5.4. *The pair $(S, \mathcal{F})$, where S is the universal set and $\mathcal{F}$ is a σ-field of subsets of S, is called a **measurable space**.*

The triple $(S, \mathcal{F}, G)$, where $(S, \mathcal{F})$ is a measurable space and G is a measure is called a **measure space**.

A **probability measure** P is a σ-finite measure defined on the measurable space $(S, \mathcal{F})$ with $P(S) = 1$.

A **probability space** $(S, \mathcal{F}, P)$ is a measure space for which P is a probability measure and $(S, \mathcal{F})$ is a measurable space.

Using the above definitions is a way of summarizing the previous two sections and introducing a very important and widely used notation: **the probability triple** $(S, \mathcal{F}, P)$. The experimental outcome space S is the set of all possible outcomes. The event space $\mathcal{F}$ is a σ-field of subsets of S. The probability measure P assigns a number (called a probability) to each event in the event space. By insisting that the event space be a σ-field we are ensuring that any sequence of set operations on a set in $\mathcal{F}$ will yield another member of $\mathcal{F}$ for which a probability has either been assigned or can be determined. The above definition of the probability triple is consistent with the axioms of probability and provides the needed structure for the event space.

We now have at our disposal a very powerful basis for applying the theory of probability. Events can be combined or otherwise operated on (usually to generate a partition of the event into "simpler" pieces), and the axioms of probability can be applied to compute (or bound) event probabilities. An extremely important conclusion is that we are always interested in (at most) a countable collection of events and the probabilities of these events. One need not be concerned with assigning a probability to each possible subset of the outcome space.

Let $(S, \mathcal{F}, P)$ be a probability space. For any $B \in \mathcal{F}$ we define

$$\int_B dP(\zeta) = P(B). \tag{1.60}$$

If $A_1, A_2, \ldots$ is a partition of B (with each $A_i \in \mathcal{F}$) then

$$P(B) = \sum_{i=1}^{\infty} P(A_i) = \sum_{i=1}^{\infty} \int_{A_i} dP(\zeta). \tag{1.61}$$

The integrals above are known as Lebesgue-Stieltjes integrals. Although a thorough discussion of integration theory is well beyond the scope of this text, the above expressions will prove useful for evaluating probabilities and providing a concise notation. The point here is that if we can compute $P(A_i)$ for all A_is in a partition of B, then we can compute $P(B)$—and hence we can evaluate the integral

$$\int_B dP(\zeta) = P(B).$$

Whether or not B is discrete, a discrete collection of disjoint (mutually exclusive) events $\{A_i\}$ can always be found to evaluate the Lebesgue-Stieltjes integrals we shall encounter. The above integral expressions also illustrate one recurring theme in our application of probability theory. To compute an event probability, partition the event into pieces (with the probability of each piece either known or "easily" computed) then sum the probabilities.

Drill Problem 1.5.1. *A pentahedral die (with faces labeled 1, 2, 3, 4, 5) is loaded so that $P(\{k\}) = kP(\{1\})$, $k = 1, 2, 3, 4, 5$. Event $A = \{1, 2, 4\}$, and event $B = \{2, 3, 5\}$. Find*

$$\int_A dP(\zeta), \quad \int_B dP(\zeta), \text{ and } \quad \int_{A \cap B} dP(\zeta).$$

Answers: 2/15, 7/15, 2/3.

Drill Problem 1.5.2. *Let $S = [0, 1]$, $A_1 = [0, 0.5]$, $A_2 = (0.45, 0.7)$, and $A_3 = \{0.2, 0.5\}$. Assume $P(\zeta \in I) =$length of the interval $I \cap S$. Find*

$$\int_{A_1 \cup A_2} dP(\zeta), \quad \int_{A_3} dP(\zeta), \text{ and } \quad \int_{A_1 \cup A_2} dP(\zeta).$$

Answers: 0, 0.05, 0.7.

1.6   INDEPENDENCE

In many practical problems in probability theory, the concept of independence is crucial to any reasonable solution. Essentially, two events are independent if the occurrence of one of the events tells us nothing about the occurrence of the other event. For example, consider a fair coin tossed twice. The outcome of a head on the first toss gives no new information concerning the outcome of a head on the second toss; the events are independent. Independence implies that the occurrence of one of the events has no effect on the probability of the other event.

Definition 1.6.1. *The two events A and B are **independent** if and only if*

$$P(A \cap B) = P(A)P(B).$$

We will find that in many problems, the assumption of independence dramatically reduces the amount of work necessary for a solution. However, independence is used only after we have verified the events are independent. The only way to test for independence is to apply the definition: if the product of the probabilities of the two events equals the probability of their intersection, then the events are independent.

Example 1.6.1. *A biased four-sided die, with faces labeled 1, 2, 3 and 4, is tossed once. If the number which appears is odd, then the die is tossed again. The die is biased in such a way that the probability of a particular face is proportional to the number on that face. Let event A be an odd number on the first toss, and event B be an odd number on the second toss. Are events A and B independent?*

Solution.   From the given information, Table 1.1 is easily filled in. The $\checkmark$ denotes that the outcome in that row belongs to the event at the top of the column.

TABLE 1.1: Summary of Example 1.6.1					
TOSS 1	TOSS 2	P(.)	A	B	A ∩ B
1	1	1/100	✓	✓	✓
1	2	2/100	✓		
1	3	3/100	✓	✓	✓
1	4	4/100	✓		
2		20/100			
3	1	3/100	✓	✓	✓
3	2	6/100	✓		
3	3	9/100	✓	✓	✓
3	4	12/100	✓		
4		40/100			

From Table 1.1 we obtain $P(A) = 0.4$ and $P(B) = P(A \cap B) = 0.16$. Since

$$P(A \cap B) = 0.16 \neq P(A)P(B),$$

the events A and B are not independent.   ∎

Many students are often confused by the relationship between independent and mutually exclusive events. Generally, two mutually exclusive events can not be independent events since the occurrence of one of the events implies that the other did not occur.

Theorem 1.6.1. *Mutually exclusive events A and B are independent iff (if and only if) either* $P(A) = 0$ *or* $P(B) = 0$.

Proof. Since A and B are mutually exclusive, we have $A \cap B = \varnothing$ so that $P(A \cap B) = P(\varnothing) = 0$. Hence $P(A \cap B) = P(A)P(B)$ iff $P(A)P(B) = 0$.   ∎

The definition of independence can be expanded when more than two events are involved.

Definition 1.6.2. *Events $A_1, A_2, \ldots, A_n$ are independent iff (if and only if)*

$$P(A_{k_1} \cap A_{k_2} \cap \cdots \cap A_{k_r}) = P(A_{k_1})P(A_{k_2}) \cdots P(A_{k_r})$$

where $k_1, k_2, \ldots, k_r$ take on every possible combination of integer values taken from $\{1, 2, \ldots, n\}$ for every $r = 2, 3, \ldots, n$.

Pairwise independence is a necessary but not a sufficient condition for independence of n events. To illustrate this definition of independence, consider the conditions that are required to have three independent events. The events A_1, A_2, and A_3 are independent if and only if

$$P(A_1 \cap A_2 \cap A_3) = P(A_1)P(A_2)P(A_3),$$

$$P(A_1 \cap A_2) = P(A_1)P(A_2),$$

$$P(A_1 \cap A_3) = P(A_1)P(A_3),$$

and

$$P(A_2 \cap A_3) = P(A_2)P(A_3).$$

The number of conditions, say N, that are necessary to establish independence of n events is found by summing all possible event combinations

$$N = \sum_{k=2}^{n} \binom{n}{k}$$

From the Binomial Theorem we have

$$(1+1)^n = \sum_{k=0}^{n} \binom{n}{k} = 1 + n + N = 2^n;$$

hence the total number of conditions is $N = 2^n - n - 1$, for $n \geq 2$.

Theorem 1.6.2. *Suppose event A can be expressed in terms of the events $A_1, A_2, \ldots, A_m$, and the event B can be expressed in terms of the events $B_1, B_2, \ldots, B_n$. If the collections of events $\{A_i\}_{i=1}^{m}$ and $\{B_i\}_{i=1}^{n}$ are independent of each other, i.e., if*

$$P(A_{k_1} \cap A_{k_2} \cap \cdots \cap A_{k_q} \cap B_{\ell_1} \cap B_{\ell_2} \cap \cdots \cap B_{\ell_r})$$
$$= P(A_{k_1} \cap A_{k_2} \cap \cdots \cap A_{k_q}) P(B_{\ell_1} \cap B_{\ell_2} \cap \cdots \cap B_{\ell_r})$$

for all possible combinations of k_is and ℓ_js, then the events A and B are independent.

Proof. Let $\{C_i\}$ be a partition of the event A and let $\{D_i\}$ be a partition of the event B. Then

$$P(A \cap B) = \sum_i \sum_j P(C_i \cap D_j) = \sum_i P(C_i) \sum_j P(D_j);$$

hence, $P(A \cap B) = P(A)P(B)$.   ∎

Example 1.6.2. *In the circuit shown in Fig. 1.10, switches operate independently of one another, with each switch having a probability of being closed equal to p. After monitoring the circuit over a*

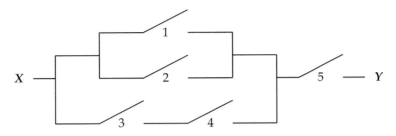

FIGURE 1.10: Circuit for Example 1.6.2

long period of time, it is observed that there is a closed path between X and Y 16.623% of the time. Find p.

Solution. Let C_i be the event that switch i is closed. A description of the circuit is then given by

$$P(A \cap C_5) = 0.16623,$$

where $A = C_1 \cup C_2 \cup (C_3 \cap C_4)$. Since A and C_5 are independent,

$$P(A \cap C_5) = P(A)P(C_5) = p\,P(A) = 0.16623.$$

With $B = C_1 \cup C_2$ and $D = C_3 \cap C_4$ we have

$$P(A) = P(B \cup D) = P(B) + P(D) - P(B \cap D),$$
$$P(B) = P(C_1) + P(C_2) - P(C_1 \cap C_2) = 2p - p^2,$$

and

$$P(D) = p^2.$$

Since B and D are independent,

$$P(B \cap D) = P(B)P(D) = (2p - p^2)p^2,$$

so that

$$P(A) = (2p - p^2)(1 - p^2) + p^2 = p^4 - 2p^3 + 2p$$

and

$$P(C) = p\,P(A) = p^5 - 2p^4 + 2p^2 = 0.16623.$$

Iterative solution yields $p = 0.3$.

Drill Problem 1.6.1. *A coin is tossed three times. The coin is biased so that a tail is twice as likely to occur as a head. Let A equal the event that two heads and one tail occur and B equal the event that more heads than tails occur. Are events A and B independent?*
Answer: No.

1.7 JOINT PROBABILITY

In this section, we introduce some notation which is useful for describing combined experiments. We have seen a number of examples of experiments which can be considered as a sequence of subexperiments—drawing five cards from a deck, for example.

Consider an experiment ε consisting of a combination of the n subexperiments ε_i, $i = 1, 2, \ldots, n$. We denote this combined experiment by the cartesian product:

$$\varepsilon = \varepsilon_1 \times \varepsilon_2 \times \cdots \times \varepsilon_n. \tag{1.62}$$

With S_i denoting the outcome space for ε_i, we denote the outcome space S for the combined experiment by

$$S = S_1 \times S_2 \times \cdots \times S_n; \tag{1.63}$$

hence, when the outcome of ε_i is $\zeta_i \in S_i$, $i = 1, 2, \ldots, n$, the outcome of ε is

$$\zeta = (\zeta_1, \zeta_2, \ldots, \zeta_n) \in S. \tag{1.64}$$

The probability that event A_1 occurs in ε_1 and event A_2 occurs in $\varepsilon_2, \ldots,$ and A_n occurs in ε_n is called the joint probability of $A_1, A_2, \ldots,$ and A_n; this joint probability is denoted by

$$P(A) = P(A_1, A_2, \ldots, A_n), \tag{1.65}$$

where $A = A_1 \times A_2 \times \cdots \times A_n$. Note that event A is the (ordered) sequence event discussed in Section 1.2.3. Let $(S_i, \mathcal{F}_i, P_i)$ denote the probability space for ε_i, and let $(S, \mathcal{F}, P)$ denote the probability space for ε. Note that the event space for ε is $F = \mathcal{F}_1 \times \mathcal{F}_2 \times \cdots \times \mathcal{F}_n$. Letting

$$A_i' = S_1 \times \cdots \times S_{i-1} \times A_i \times S_{i+1} \times \cdots \times S_n, \tag{1.66}$$

we find that

$$P(A) = P(A_1' \cap A_2' \cap \cdots \cap A_n'). \tag{1.67}$$

In particular, we may find $P_i(A_i)$ from $P(\cdot)$ using

$$P_i(A_i) = P(A_i') = P(S_1, \ldots, S_{i-1}, A_i, S_{i+1}, \ldots, S_n). \tag{1.68}$$

We sometimes (as in the previous examples) simply write

$$P(A) = P(A_1 A_2 \cdots A_n) \qquad (1.69)$$

for the probability that the sequence event $A = A_1 A_2 \cdots A_n$ occurs. We also sometimes abuse notation and treat $P(A_1)$, $P_1(A_1)$, and $P(A_1')$ as identical expressions.

It is important to note that, in general, we cannot obtain $P(\cdot)$ from $P_1(\cdot)$, $P_2(\cdot)$, ..., and $P_n(\cdot)$. An important exception is when the experiments ε_1, ε_2, $\cdots$, ε_n are independent.

Definition 1.7.1. *The experiments ε_1, ε_2, $\cdots$, ε_n are **independent** iff*

$$P(A) = P_1(A_1) P_2(A_2) \cdots P_n(A_n) \qquad (1.70)$$

for all $A_i \in \mathcal{F}_i$, $i = 1, 2, \ldots, n$.

Example 1.7.1. *A combined experiment $\varepsilon = \varepsilon_1 \times \varepsilon_2$ consists of drawing two cards from an ordinary deck of 52 cards. Are the subexperiments independent?*

Solution. Consider the drawing of two hearts. We have

$$P(H_1 H_2) = \frac{13 \times 12}{52 \times 51} = \frac{1}{17},$$

$$P_1(H_1) = \frac{13}{52} = \frac{1}{4},$$

and

$$P_2(H_2) = P(H_1 H_2) + P(X H_2) = \frac{1}{17} + \frac{39 \times 13}{52 \times 51} = \frac{1}{4},$$

where X consists of the 39 non-heart cards. Hence, the subexperiments ε_1 and ε_2 are not independent since

$$P(H_1 H_2) = \frac{1}{17} \neq \frac{1}{16} = P_1(H_1) P_2(H_2). \qquad \blacksquare$$

Drill Problem 1.7.1. *A combined experiment $\varepsilon = \varepsilon_1 \times \varepsilon_2$ consists of drawing two cards from an ordinary deck of 52 cards. Let H_1 denote the drawing of a heart on the first draw, and H_2 denote the drawing of a heart on the second draw. Let $H_1 = H_1 \times S_2$ and $H_2' = S_1 \times H_2$. Find $P(H_1')$, $P(H_2')$, and $P(H_2' \cap H_1')$.*

Answers: 1/17, 1/4, 1/4.

1.8   CONDITIONAL PROBABILITY

Assume we perform an experiment and the result is an outcome in event A; that is, additional information is available but the exact outcome is unknown. Since the outcome is an element of event A, the chances of each sample point in event A occurring have improved and those in event A^c occurring are zero. To determine the increased likelihood of occurrence for outcomes in event A due to the additional information about the result of the experiment, we scale or correct the probability of all outcomes in A by $\frac{1}{P(A)}$.

Definition 1.8.1. *The **conditional probability** of an event B occurring, given that event A occurred, is defined as*

$$P(B|A) = \frac{P(A \cap B)}{P(A)}, \tag{1.71}$$

provided that $P(A)$ is nonzero.

Note carefully that $P(B|A) \neq P(A|B)$. In fact, we have

$$P(A|B) = \frac{P(A \cap B)}{P(B)} = \frac{P(B|A)P(A)}{P(B)}. \tag{1.72}$$

The vertical bar in the previous equations should be read as "given," that is, the symbol $P(B|A)$ is read "the probability of B, given A." The conditional probability measure is a legitimate probability measure that satisfies each of the axioms of probability.

Using the definition of conditional probability, the events A and B are independent if and only if

$$P(A|B) = \frac{P(A \cap B)}{P(B)} = \frac{P(A)P(B)}{P(B)} = P(A).$$

Similarly, A and B are independent if and only if $P(B|A) = P(B)$. Each of the latter conditions can be (and often is) taken as an alternative definition of independence. The one difficulty with this is the case where either $P(A) = 0$ or $P(B) = 0$. If $P(B) = 0$, we can define $P(A|B) = P(A)$; similarly, if $P(A) = 0$, we can define $P(B|A) = P(B)$.

Conditional probabilities, given event $A \in \mathcal{F}$, on the probability space $(S, \mathcal{F}, P)$ can be treated as unconditional probabilities on the probability space $(S_A, \mathcal{F}_A, P_A)$, where $S_A = S \cap A = A$, $\mathcal{F}_A$ is a σ-field of subsets of S_A, and P_A is a probability measure. The σ-field is $\mathcal{F}_A$ the restriction of $\mathcal{F}$ to A defined by

$$\mathcal{F}_A = \{A \cap B : B \in \mathcal{F}\}.$$

The proof that $\mathcal{F}_A$ is indeed a σ-field of subsets of S_A is left to the reader (see Problem 1.45). The probability measure P_A is defined on the measurable space $(S_A, \mathcal{F}_A)$ by

$$P_A(B_A) = \frac{P(B \cap A)}{P(A)} = P(B|A). \qquad (1.73)$$

If $P(A) = 0$ we may define P_A to be any valid probability measure; in this case, $P_A(B_A) = P(B)$ is often a convenient choice.

Now consider the conditional independence of events A and B, given an event C occurred. Above, we interpreted a conditional probability as an unconditional probability defined with a sample space equal to the given event. Thus, the conditional independence of A and B is established in the new sample space C by testing

$$P_C(A_C \cap B_C) = P_C(A_C)P_C(B_C)$$

or in the original sample space by testing

$$P(A \cap B|C) = P(A|C)P(B|C).$$

Note that independent events are not necessarily conditionally independent (given an arbitrary event).

Example 1.8.1. *A number is drawn at random from $S = \{1, 2, \ldots, 8\}$. Define the events $A = \{1, 2, 3, 4\}$, $B = \{3, 4, 5, 6\}$, and $C = \{3, 4, 5, 6, 7, 8\}$. (a) Are A and B independent? (b) Are A and B independent, given C?*

Solution. (a) We find $P(A) = P(B) = 1/2$ and $P(A \cap B) = 1/4 = P(A)P(B)$, so that A and B are independent. (b) We find $P(A|C) = 1/3$, $P(B|C) = 2/3$, and $P(A \cap B|C) = 1/3$, so that A and B are not independent, given C.

Random Sampling of Waveform

One is often interested in studying the frequency of occurrence of certain events even when the observed phenomena is inherently deterministic (not random). Here, we consider the "random" sampling of a deterministic waveform. Sample the time function $f(t)$ uniformly at $t = kT$, for $k = 0, 1, \ldots, N - 1$, where $T > 0$ is called the sampling period. The outcome space for the experiment is the set of all pairs of k and $f(kT)$: $S = \{(k, f(kT)) : k = 0, 1, \ldots, N - 1\}$. We assume that each outcome in the outcome space is equally likely to occur.

Example 1.8.2. *The waveform shown in Fig. 1.11 is uniformly sampled every one-half second from $(T = 0.5)$. Define the events $A = \{f(kT) > 5/4\}$, $B = \{0.5 \le f(kT) < 2\}$, and $C = \{2 \le t < 3\}$. Find (a) $P(A|B)$, (b) $P(C)$, (c) $P(B|C)$.*

Solution. There are nine equally likely outcomes. In the following list, a check in the appropriate row indicates whether the sample point ζ is an element of event A, B, or C.

SAMPLE POINT, ζ	A	B	C
(0,0)			
(1,0.5)		✓	
(2,1)		✓	
(3,1.5)	✓	✓	
(4,2)	✓		✓
(5,1.5)	✓	✓	✓
(6,1)		✓	
(7,0.5)		✓	
(8,0)			

(a) With the aid of the above list, we find $P(A) = 3/9$, $P(B) = 6/9$, and $P(A \cap B) = 2/9$. Hence

$$P(A \mid B) = \frac{P(A \cap B)}{P(B)} = \frac{2/9}{6/9} = \frac{1}{3}.$$

(b) We find $P(C) = 2/9$.

(c) We have $P(B \cap C) = 1/9$, so that

$$P(B|C) = \frac{P(B \cap C)}{P(C)} = \frac{1/9}{2/9} = \frac{1}{2}. \qquad \blacksquare$$

Probability Tree

A probability tree is a natural extension of both the concept of conditional probability and the tree diagram. The tree diagram is drawn with the tacit understanding that the event on the right side of any branch occurs, given that the sequence event on the pathway from the origin to the left side of the branch occurred. On each branch, we write the conditional probability of the

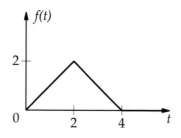

FIGURE 1.11: Waveform for Example 1.8.2

event at the node on the right side of the branch, given the sequence event on the pathway from the origin to the node on the left side of the branch. The probabilities for the branches leaving the origin node are, of course, written as unconditioned probabilities (note that $P(A|S) = P(A)$). The probability of each event equals the product of all the branch probabilities connected from the origin to its terminal node. A tree diagram with a probability assigned to each branch is a probability tree. The following theorem and its corollary justify the technique.

Theorem 1.8.1. *For arbitrary events $A_1, A_2, \ldots, A_n$ we have*

$$P(A_1 \cap A_2 \cap \cdots \cap A_n) \qquad\qquad (1.74)$$
$$= P(A_1)P(A_2|A_1)P(A_3|A_1 \cap A_2) \cdots P(A_n|A_1 \cap A_2 \cap \cdots \cap A_{n-1}).$$

Proof. For any $k \in \{2, 3, \ldots, n\}$, we have

$$P(A_1 \cap A_2 \cdots \cap A_k) = P(A_k|A_1 \cap A_2 \cdots \cap A_{k-1})P(A_1 \cap A_2 \cdots \cap A_{k-1});$$

applying this for $k = n, n-1, \ldots, 2$ establishes the desired result. ∎

The above theorem provides a useful expansion for the probability of the intersection of n events in terms of conditional probabilities. An intersection of n events is not an ordered event; however, the treatment of joint probabilities in the previous section enables us to apply the above theorem to an (ordered) sequence event and establish the following corollary.

Corollary 1.8.1. *The probability for the sequence event $A_1 A_2 \cdots A_n$ may be expressed as*

$$P(A_1 A_2 \cdots A_n) = P(A_1)P(A_2|A_1)P(A_3|A_1 A_2) \cdots P(A_n|A_1 A_2 \cdots A_{n-1}). \qquad (1.75)$$

Example 1.8.3. *Two cards are drawn at random from an ordinary deck of 52 cards without replacement. Find the probability p that both are spades.*

Solution. Let us set up a probability tree with event S_i denoting a spade drawn on the ith draw, as shown in Fig. 1.12. The probability of any event in the probability tree is equal to the product of all of the conditional probabilities of the branches connected on the pathway from the origin to the left of the event. Therefore

$$p = P(S_1 S_2) = P(S_1)P(S_2|S_1) = \frac{13}{52}\frac{12}{51} = \frac{1}{17}.$$

Partitioning a sample space often simplifies a problem solution. The Theorem of Total Probability provides analytical insight into this important process.

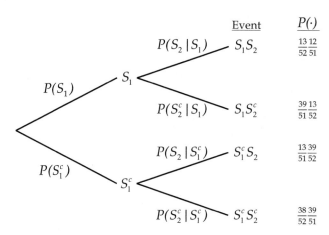

FIGURE 1.12: Probability tree for Example 1.8.3

Theorem 1.8.2 (Total Probability). *Let A_1, A_2, ..., A_n be a partition either of S or of B. Then*

$$P(B) = \sum_{i=1}^{n} P(B|A_i)P(A_i). \tag{1.76}$$

Proof. We have

$$B = B \cap (A_1 \cup A_2 \cup \cdots \cup A_n)$$

so that

$$P(B) = \sum_{i=1}^{n} P(B \cap A_i) = \sum_{i=1}^{n} P(B|A_i)P(A_i). \qquad \blacksquare$$

Partitioning a sample space is usually logical and can result in a solution of what appears at first to be an extremely difficult problem. In fact, the Theorem of Total Probability allows us to easily solve problems that have previously been solved by using a probability tree, as in the next example.

Example 1.8.4. *We have four boxes with a composition of defective light bulbs as follows: Box B_i contains 5%, 40%, 10%, and 25% defective light bulbs for $i = 1, 2, 3,$ and 4, respectively. Pick a box and then pick a light bulb from that box at random. What is the probability that the light bulb is defective?*

Solution. We solve this problem first using a probability tree and then by applying the Theorem of Total Probability. Since each box is equally likely, we have $P(B_i) = 1/4$ for $i = 1, 2, 3, 4$. Let D be the event that a defective light bulb is selected. From the probability tree shown in

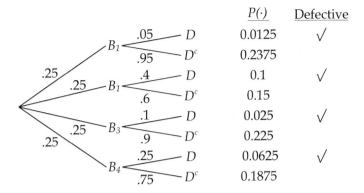

FIGURE 1.13: Probability tree for Example 1.8.4

Fig. 1.13 we have

$$P(D) = \sum_{i=1}^{4} P(B_i \cap D) = 0.0125 + 0.1 + 0.025 + 0.0625 = 0.2.$$

From the Theorem of Total Probability,

$$P(D) = \sum_{i=1}^{4} P(B_i)P(D|B_i) = \frac{1}{4}(0.05 + 0.4 + 0.1 + 0.25) = 0.2. \qquad \blacksquare$$

A Posteriori Probabilities

The probabilities that have been computed in all of our previous examples are known as *a priori* probabilities. That is, we are computing the probability of some event that may or may not occur in the future. Time is introduced artificially, but it allows us to logically follow a sequence. Now, suppose that some event has occurred and that our observation of the event has been imperfect. Given our imperfect observation, can we deduce a conditional probability for this event having occurred? The answer is yes, such a probability is known as an *a posteriori* probability and Bayes' Theorem provides a method of computing it.

Theorem 1.8.3 (Baye's Theorem). *Let $A_1, A_2, \ldots, A_n$ be a partition of the outcome space S, and let $B \in \mathcal{F}$ be an arbitrary event. Then*

$$P(A_i|B) = \frac{P(A_i)P(B|A_i)}{\sum_{j=1}^{n} P(A_j)P(B|A_j)}. \qquad (1.77)$$

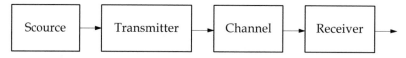

FIGURE 1.14: Typical communication system

Proof.   From the definition of conditional probability,

$$P(A_i|B) = \frac{P(A_i \cap B)}{P(B)} = \frac{P(A_i)P(B|A_i)}{P(B)}.$$

Using the Theorem of Total Probability to express $P(B)$ yields the desired result.   ∎

Example 1.8.5.   *In Example 1.8.4, suppose the light bulb was defective. What is the probability it came from Box 2?*

Solution.   From Bayes' Theorem

$$P(B_2|D) = \frac{P(B_2)P(D|B_2)}{P(D)} = \frac{0.1}{0.2} = 0.5.$$

Example 1.8.6.   *The basic components of a binary digital communication system are shown in Fig. 1.14. Every T seconds, the source puts out a binary digit (a one or a zero), which is transmitted over a channel to the receiver. The channel is typically a telephone line, a fiber optic cable, or a radio link, subject to noise which causes errors in the received digital sequence, a one is interpreted as a zero and vice versa. Let us include the uncertainty introduced due to noise in the channel for any period with the probability tree description shown in Fig. 1.15, where S_i is the binary digit i sent by the source, and R_i is the binary digit i captured by the receiver.*

Determine (a) the probability of event C, the signals are received with no error, and (b) which binary digit has the greater probability of being sent, given the signal was received correctly.

$$P(S_0) = 0.7 \quad S_0 \quad \begin{array}{l} P(R_0|S_0) = 0.8 \quad R_0 \\ P(R_1|S_0) = 0.2 \quad R_1 \end{array}$$

$$P(S_1) = 0.3 \quad S_1 \quad \begin{array}{l} P(R_0|S_1) = 0.1 \quad R_0 \\ P(R_1|S_1) = 0.9 \quad R_1 \end{array}$$

FIGURE 1.15: Probability tree for Example 1.8.6

Solution

(a) From the Theorem of Total Probability

$$P(C) = P(S_0)P(R_0|S_0) + P(S_1)P(R_1|S_1) = \frac{7}{10}\frac{8}{10} + \frac{3}{10}\frac{9}{10} = 0.83.$$

(b) From Bayes' Theorem, we determine

$$P(S_0|C) = \frac{P(S_0)P(C|S_0)}{P(C)} = \frac{(7/10)(8/10)}{83/100} = \frac{56}{83}$$

and

$$P(S_1|C) = \frac{P(S_1)P(C|S_1)}{P(C)} = \frac{(3/10)(9/10)}{83/100} = \frac{27}{83}$$

which implies that a zero has the greater probability of being sent correctly. ∎

Drill Problem 1.8.1. *The waveform $f(t)$ is uniformly sampled every 0.1s from 0 to 4 s, where*

$$f(t) = \begin{cases} 2t & \text{if } 0 \leq t \leq 2 \\ 4e^{-2(t-2)} & \text{if } 2 < t \leq 4. \end{cases}$$

Evaluate the probability that: (a) $f(t) \leq 1$; (b) $f(t) \geq 2$, given $f(t) \leq 3$; (c) a value is from the $f(t) = 2t$ portion of the curve, given $f(t) \leq 1$.

Answers: 8/35, 20/41, 6/20.

Drill Problem 1.8.2. *Two balls are drawn without replacement from an urn that contains four green, six blue, and two white balls. Evaluate the probability that: (a) both balls are white; (b) one ball is white and one ball is green; (c) the first ball is blue, given that the second ball is green.*

Answers: 24/44, 16/132, 2/132.

Drill Problem 1.8.3. *The waveform $f(t)$ is uniformly sampled every 0.1 s from -1 to 2 s, where*

$$f(t) = \begin{cases} -t & \text{if } -1 \leq t \leq 0 \\ t & \text{if } 0 < t \leq 1 \\ \sin\left(\frac{1}{2}\pi t\right) & \text{if } 1 < t \leq 2. \end{cases}$$

Evaluate the probability that: (a) $f(t) \leq 0.5$, given $-1 \leq t \leq 0$; (b) $f(t) \leq 0.5$, given $0 < t \leq 1$; (c) $f(t) \leq 0.5$, given $1 < t \leq 2$; (d) $f(t) \leq 0.5$; (e) $-1 \leq t \leq 0$, given $f(t) \leq 0.5$.

Answers: 6/15, 1/2, 6/11, 4/10, 15/31.

Drill Problem 1.8.4.   *Four boxes contain the following quantity of marbles.*

	RED	BLUE	GREEN
Box 1	6	3	2
Box 2	5	4	0
Box 3	3	3	4
Box 4	2	9	7

A box is selected at random and the marble selected is green. Determine the probability that: (a) box 1 was selected, (b) box 2 was selected, (c) box 3 was selected, (d) box 4 was selected.

Answers: 0, 385/961, 180/961, 396/961.

1.9   SUMMARY

In this chapter, we have studied the fundamentals of probability theory upon which all of our future work is based. Our discussion began with the preliminary topics of set theory, the sample space for an experiment, and combinatorial mathematics. Using set theory notation, we developed a theory of probability which is summarized by the probability space $(S, \mathcal{F}, P)$, where S is the experimental outcome space, $\mathcal{F}$ is a σ-field of subsets of S ($\mathcal{F}$ is the event space), and P is a probability measure which assigns a probability to each event in the event space. It is important to emphasize that the axioms of probability do not dictate the choice of probability measure P. Rather, they provide conditions that the probability measure must satisfy. For the countable outcome spaces that we have seen so far, the probabilities are assigned using either the classical or the relative frequency method.

Notation for joint probabilities has been defined. The concept of joint probability is useful for studying combined experiments. Joint probabilities may always be defined in terms of intersections of events.

We defined two events A and B to be independent iff (if and only if)

$$P(A \cap B) = P(A)P(B).$$

The extension to multiple events was found to be straightforward.

Next, we introduced the definition for conditional probability as the probability of event B, given event A occurred

$$P(B|A) = \frac{P(A \cap B)}{P(A)},$$

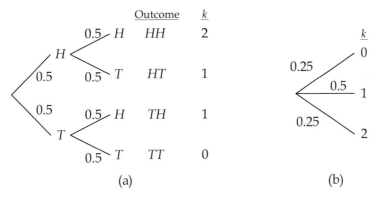

FIGURE 1.16: Partial tree diagram for Example 1.9.1

provided that $P(A) \neq 0$. The extension of the definition of conditional probability to multiple events involved no new concepts, just application of the axioms of probability. We next presented the Theorem of Total Probability and Bayes' Theorem.

Thus, this chapter presented the basic concepts of probability theory and illustrated techniques for solving problems involving a countable outcome space. The solution, as we have seen, typically involves the following steps:

1. List or otherwise describe events of interest in an event space $\mathcal{F}$,

2. Assignment/computation of probabilities, and

3. Solve for the desired event probability.

The following example illustrates each of these steps in the solution.

Example 1.9.1. *An experiment begins by rolling a fair tetrahedral die with faces labeled 0, 1, 2, and 3. The outcome of this roll determines the number of times a fair coin is to be flipped.*

(a) *Set up a probability tree for the event space associating the outcome of the die toss and the number of heads flipped.*

(b) *If there were two heads tossed, then what is the probability of a 2 resulting from the die toss?*

Solution. Let n be the value of the die throw, and k be the total number of heads resulting from the coin flips.

(a) Since it is fairly difficult to draw the probability tree for this experiment directly, we shall develop it in stages. We first draw a partial probability tree shown in Fig. 1.16(a) for the case in which the die outcome is two and the coin is flipped twice. This probability tree can be compressed into a more efficient event space representation as shown in

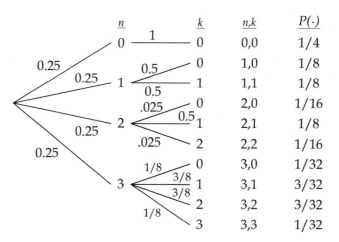

FIGURE 1.17: Tree diagram for Example 1.9.1

Fig. 1.16(b). It should be clear that the probability of any face occurring on the die toss is equal to 1/4 and that the values that k takes on as well as the probabilities are dependent on n.

We can draw the probability tree for the entire event space by continuing in the same manner as before. The result is shown in Fig. 1.17.

(b) Using Bayes' Theorem,

$$P(n = 2|k = 2) = \frac{P(n = 2)P(k = 2|n = 2)}{P(k = 2)}.$$

To find $P(k = 2)$ we can use the tree diagram and sum the probabilities for all events that have $k = 2$ or use the Theorem of Total Probability to obtain

$$P(k = 2) = P(k = 2|n = 2)P(n = 2) + P(k = 2|n = 3)P(n = 3),$$

so that $P(k = 2) = \frac{1}{16} + \frac{3}{32} = \frac{5}{32}$. Finally,

$$P(n = 2|k = 2) = \frac{(1/4)(1/4)}{5/32} = \frac{2}{5}.$$

Drill Problem 1.9.1. *Professor S. Rensselaer teaches a course in probability theory. She is a kind-hearted but very tricky old lady who likes to give many unannounced quizzes each week. She determines the number of quizzes each week by tossing a fair tetrahedral die with faces labeled 0, 1, 2, and 3. The more quizzes she gives, however, the less time she has to assign and grade homework problems. If Ms. Rensselaer is to give Q quizzes during the next week, then the conditional probability she will assign*

H homework problems is given by

$$P(H|Q) = \begin{cases} \frac{1}{4-Q}, & \text{if } 1 \le H \le 4 - Q \\ 0 & \text{otherwise.} \end{cases}$$

Determine the probability: (a) that two homework problems are assigned during the week; (b) of one quiz during the week that two homework problems were assigned; (c) that two homework problems are assigned, given at least one quiz during the week.

Answers: 5/18, 13/48, 4/13.

1.10   PROBLEMS

1.1. Let $A \cup (A^c \cup B^c)^c = \{1, 5, 8\}$, $A \cup (A^c \cap B) \cup (A \cap B \cap C) = \{1, 2, 5, 8, 9\}$, and $(C^c \cup A^c)^c \cup (C^c \cup A)^c = \{1, 5, 7\}$. Furthermore, A and B are mutually exclusive, and A, B, and C, are collectively exhaustive. Determine: (a) A, (b) B, (c) C^c, (d) S.

1.2. Given a sample space $S = \{-1.0, -0.9, -0.8, \ldots, 1.8, 1.9, 2.0\}$, and the following six sets in the sample space:

$$A = \{x : -1 \le x \le 0\}, \quad B = \{x : -x \le 1/2\},$$
$$C = \{x : 0 < x \le 1\}, \quad D = \{x : x \le 1/2\},$$
$$E = \{x : 1 < x \le 2\}, \quad F = \{x : \sin(\pi x/2) \le 1/2\}.$$

Note that all elements of A, B, etc. must also be in S. Find: (a) $A \cap B$, (b) $C \cap D$, (c) $E \cap F$, (d) $(A \cap B) \cup (A \cap B^c) \cup (A^c \cup B)$, (e) $A \cup ((A^c \cup C^c)^c \cap B)$.

1.3. Let the sample space be all real numbers, and define the sets $A = \{x : x > 0\}$, $B = \{x : x/2 \text{ is an integer}\}$ and $C = \{x : 2 < x < 8\}$ be defined on the sample space. Find: (a) $(A \cap C) \cup B$, (b) $A^c \cap B$, (c) $A \cap C^c$, (d) $A \cap B \cap C$, (e) $A \cup B \cup C$.

1.4. Prove that if $B \subset A$ then (a) $A \cup B = A$, and (b) $A \cap B = B$.

1.5. Simplify: (a) $(A \cap B) \cup (A \cap B^c)$, (b) $(D^c \cap C^c) \cup (C \cap D^c)$, (c) $(A \cap B) \cap (B^c \cap A)$, (d) $D^c \cap C^c \cap D^c \cap C$.

1.6. Prove or give a counterexample:

(a) If $A \cup B = A \cup C$ then $B = C$.

(b) If $A \cap B = A \cap C$ then $B = C$.

1.7. Prove that if $A \cap B = \varnothing$, $A \cap C = \varnothing$, and $A \cup B = A \cup C$, then $B = C$.

1.8. Prove the following identities for arbitrary sets A, B, and C:

(a) $A = (A \cap B) \cup (A \cap B^c)$,
(b) $A \cup (A^c \cap B) = B \cup (A \cap B^c)$,
(c) $(A \cup B) \cap (A \cup B^c) = A$,
(d) $(A \cup B) \cap (A \cap B)^c = (A \cap B^c) \cup (B \cap A^c)$,
(e) $(A \cap B) \cup ((A \cap B) \cup (A^c \cap B^c))^c = A \cup B$,
(f) $(A \cap B \cap C^c) \cup ((A \cap C) \cup (A^c \cap B \cap C) \cup (B^c \cap C))^c = C^c$.

1.9. Determine the validity of the following relationships for arbitrary sets A, B, and C.

(a) $(A^c \cap (C \cup B)^c) \cup (A^c \cap B) \cup ((A \cup (A^c \cap B))^c \cap C) = A^c \cap B^c$,
(b) $(A \cup B) \cap (A \cap B)^c = (A \cap B^c) \cup (B \cap A^c) \cup (A \cap B^c \cap C)$,
(c) $(A \cup B \cup C)^c \cup (A^c \cap B^c \cap C^c) = (A^c \cup B^c \cup C^c)^c$.

1.10. If A_1, A_2, A_3 and A_4 are mutually exclusive sets and $B \subset (A_1 \cup A_2 \cup A_3 \cup A_4)$, then show that

$$B = (A_1 \cap B) \cup (A_2 \cap B) \cup (A_3 \cap B) \cup (A_4 \cap B)$$

and illustrate with a Venn diagram.

1.11. (a) The sets A and B are mutually exclusive and collectively exhaustive. Are A^c and B^c mutually exclusive? Prove it.
(b) The sets A and B are mutually exclusive but not collectively exhaustive. Are A^c and B^c mutually exclusive? Prove it.
(c) The sets A and B are mutually exclusive but not collectively exhaustive. Are A^c and B^c collectively exhaustive? Prove it.
(d) The sets A and B are collectively exhaustive but not mutually exclusive. Are A^c and B^c collectively exhaustive? Prove it.

1.12. Let $S = \{(x, y) : x \geq 0, y \geq 0\}$, $A = \{(x, y) : x + y \geq 1\}$, $B = \{(x, y) : xy < 1\}$, and $C = \{(x, y) : x < y\}$. Sketch the following sets in a coordinate space. (a) $(A \cap B)^c$, (b) $A \cap C$, (c) $A \cup (A^c \cap B)$, (d) $(A \cup (A^c \cap B))^c \cap C$.

1.13. Four boxes contain marbles labeled with numbers as shown:

	MARBLES
Box 1	1, 2, 3, 4, 5
Box 2	1, 2
Box 3	3, 5, 7
Box 4	1, 2, 4

1.14. Five playing cards, two spades, a heart, a diamond and a club, are shuffled and placed face downwards on a table. An experiment consists of drawing a card and noting its suit, then drawing another card (without replacing the first card) and noting its suit. Illustrate the sample space with a coordinate system and then with a tree diagram.

1.15. An experiment consists of tossing a coin until either three heads or two tails have appeared (not necessarily in a row). Illustrate the sample space with a tree diagram.

1.16. A transistor having three leads (an emitter, base, and collector) is connected to three points in a network. How many ways can the transistor be connected? Draw a tree diagram illustrating all possible outcomes and list the sample space.

1.17. A class of five students is given two As and three Bs. Draw a tree diagram illustrating all possible ways the grades can be assigned.

1.18. Urn 1 contains five red, two white, and three blue balls. Urn 2 contains three red and one white balls. A ball is drawn at random from urn 1 and placed in urn 2. A ball is then drawn at random from urn 2. Illustrate the sample space using a tree diagram.

1.19. An experiment of measuring resistance is performed. Find and classify the sample space.

1.20. The reactive part of an impedance is measured. Find and classify the sample space.

1.21. Let $A = \{3, 4\}$, $B = \{1, 2, 6\}$, and $C = [1, 3]$. Find $D = A \times B$, $E = B \times A$, and $F = A \times C$. Sketch D, E, and F using a coordinate system representation. Sketch a tree diagram for D and E.

1.22. Consider the letters A through F being elements of S. If each letter can only be used once, then determine the number of three letter words (a) possible, (b) possible if the letter E is second, (c) possible if a vowel must be included, (d) possible if the letters A and F (together) are included only when the letter C is present.

1.23. An experiment involves rolling three colored, six-sided dice (yellow, red, and blue). (a) What are the total number of outcomes possible? (b) How many outcomes are possible if the red die shows a three? (c) How many outcomes are possible if the red die shows an even number? (d) How many outcomes are possible if each die shows a different number?

1.24. Twelve runners, A through L, have entered a race. They are competing for first, second, and third places. Determine the number of finishes: (a) possible; (b) if runner G finishes first; (c) if runner C finishes in one of the first three places; (d) if runner D trips and does not even finish the race; (e) if one and only one of the runners, A, B, or C, finishes in one of the first three places.

1.25. In how many ways can four red, four blue and two green flags be hung (a) in a row, (b) in a row if the end flags are red?

1.26.   A class of 40 students is awarded grades of A, B, C, D, and F. Determine the number of ways the grades can be awarded if: (a) there are 5 As, 10 Bs, 15 Cs, 5 Ds, and 5 Fs; (b) all As are given; and (c) there are 20 As and 20 Bs.

1.27.   A not too bright electrical engineering student was told to solder a 16 conductor cable to a connector. Since the student lost the wiring diagram he decided to arbitrarily make connections until it worked. How many ways could he connect it?

1.28.   Five components are selected from a very large bin of components. Each component is either good or defective. How many ways can a group of five components: (a) have exactly three good components, (b) at least three good components?

1.29.   In how many ways can a set of nine distinct elements be partitioned in three groupings consisting of two, three and four elements, if the ordering in each grouping is: (a) unimportant, (b) important.

1.30.   A change purse contains five nickels, eight dimes, and three quarters. Assume each coin is distinct and the order of selection is unimportant. Determine the number of ways to select: (a) a single dime, (b) exactly 60 cents if three coins are selected, (c) exactly 60 cents if four coins are selected, (d) exactly 60 cents.

1.31.   Repeat Problem 30 if (i) the dimes, nickels and quarters, are indistinguishable, (ii) order of selection is important, (iii) order of selection is important and similar coins indistinguishable.

1.32.   A club is made up of six doctors, seven lawyers, and five plumbers. A committee of four is to be selected from the club members. (a) In how many ways can the committee be formed? (b) In how many ways can the committee be formed if at least one plumber is on the committee? (c) How many committees can be formed with exactly two doctors? (d) In how many ways can the committee be formed if one particular doctor and one particular lawyer won't serve on the committee together?

1.33.   A fair die is rolled ten times. We are only interested in the fifth face being up or not. How many ways can the fifth face be up: (a) exactly once, (b) at least once?

1.34.   Consider an ordinary deck of 52 playing cards. (a) How many different hands of five cards can be drawn? (b) How many different hands containing four aces and one other card can be formed? (c) Comment on why the hand with four aces is unlikely.

1.35.   A man has been dealt five cards from a standard 52 card deck: two spades, two hearts, and a diamond. He sets these cards aside, and is dealt five more cards. What is the probability that of these five new cards: (a) all are spades? (b) two are hearts and two are diamonds? (c) none are diamonds or spades?

1.36.   Prove that if $A \cap B = \emptyset$, then $P(A) \leq P(B^c)$.

1.37. Prove that

$$P(A \cup B \cup C) = P(A) + P(B) + P(C) - P(A \cap B) - P(A \cap C)$$
$$- P(B \cap C) + P(A \cap B \cap C).$$

1.38. Two fair dice are tossed. What is the probability that: (a) the sum of seven will appear? (b) the sum of two will appear? (c) the sum of ten will appear?

1.39. A box contains 30 transistors. Three of the transistors are known to be defective. What is the probability that four of the transistors selected at random are good?

1.40. Four runners have entered a race. Runner A is twice as likely to win as runner B. Runner C is three times as likely as B to win, and D is twice as likely as A to win. What is the probability that A or B win the race?

1.41. A four-sided die is loaded so that a one or a two is four times as likely to occur as a three or a four. The die is rolled twice. Let x denote the sum of the two rolls. Define the events A, B, and C as $A = \{x : x \text{ is even}\}$, $B = \{x : x \leq 4\}$, and $C = \{x : x > 5\}$. Determine: (a) $P(A)$, (b) $P(B)$, (c) $P(C)$, (d) $P(A \cup B)$, (e) $P(B \cup C)$, (f) $P(A^c \cup B^c)$.

1.42. What is the probability that a hand of five cards drawn at random from a standard deck of 52 cards will contain (a) the ace, king, queen, jack, and ten of clubs? (b) the ace, king, queen, jack, and ten of any one suit? (c) four of a kind? (d) a full house?

1.43. An urn contains three red and seven blue marbles. If two of the marbles are drawn at random without replacement, find the probability that (a) both are blue, (b) both are red, (c) one is red and the other is blue.

1.44. A fair die is tossed once. Event $A = \{1, 2, 3\}$ and event $B = \{2, 4, 5\}$. (a) Find the minimal σ-field containing events A and B. (b) How many elements are in the σ-field containing all subsets of S?

1.45. Let $\mathcal{F}$ be a σ-field of subsets of S. Let $A \subset S$, with A not necessarily in $\mathcal{F}$. Define

$$\mathcal{F}_A = \{A \cap E : E \in \mathcal{F}\}.$$

Show that $\mathcal{F}_A$ is a σ-field of subsets of A. The σ-field $\mathcal{F}_A$ is called the *restriction* of $\mathcal{F}$ to A.

1.46. A two bit binary word can be formed in four ways. The probability of the four single element events are $P(\{11\}) = 5/16$, $P(\{10\}) = 3/16$, $P(\{01\}) = 3/16$, and $P(\{00\}) = 5/16$. Can this experiment consist of two independent trials?

1.47. We are given that events A and B are independent. (a) Are A^c and B independent? Prove it. (b) Are A^c and B^c independent? Prove it.

1.48.   Let A and B be two events with $P(A) = 0.02$, $P(B \cap A^c) = 0.01$, and $P(A \cap B) = 0.015$. Determine: (a) if A and B are independent, (b) $P(A \cap B^c)$.

1.49.   Let A and B be two events with $P(A) = 1/8$, $P(B) = 1/4$, and $P(A \cup B) = 5/16$. Determine: (a) $P(B \cap A^c)$, (b) $P(A \cup B^c)$, (c) $P(A|B)$, (d) $P(A^c|B)$.

1.50.   County Road 7 is quite a dangerous road. The probability a driver has an accident is 0.4, a breakdown is 0.5, and neither an accident nor breakdown is 0.2. Determine: (a) the probability that the driver has a breakdown and not an accident; (b) the conditional probability that the driver has an accident, given that there is no breakdown.

1.51.   Professor Rensselaer's theory of classroom instruction is that 25% of the students in her class do not listen to her lecture, 15% of the students do not read what she writes (on the blackboard) during lecture, and 20% of the students, read what she writes and do not listen to her lecture. (a) Determine the percentage of students that have no idea what is going on in lecture. (b) Determine the probability that the student reads what is written, given the student either reads what is written or listens to the lecture.

1.52.   Fargo Polytechnic Institute (FPI) plays ten football games during a season. Let A be the event that FPI scores at least as many points as the other team, B the event that FPI wins, C the event the two teams tie, and D the event that FPI loses. Furthermore, $P(A) = 7/10$ and $P(B|D^c) = 6/7$. Determine (a) $P(B)$, (b) $P(C)$, (c) $P(B|C)$, (d) the probability FPI wins three games and loses four games, (e) the probability FPI wins at least eight games.

1.53.   A club has six members, three are men and three are women. A committee of three members is to be selected. Determine (a) the probability that any particular member is chosen, (b) the probability that at least one woman is chosen, (c) the probability that a particular man and woman are not chosen together.

1.54.   Flip a biased coin to determine which of two urns to select from. Urn A contains two white and two black balls. Urn B contains four white and one black balls. If the outcome of the coin toss is a head, select from urn A, otherwise, select from urn B. The experiment continues by drawing balls from the selected urn until a black ball is picked, at which time the game is concluded. After playing this game many times, it is observed that the game is concluded in exactly two draws from the urn with a probability of 38/150. Determine (a) the probability of selecting from urn A; (b) the probability of selecting from urn A, given two white balls are selected.

1.55. The waveform $f(t)$ is uniformly sampled every 0.1 s from $0 \le t \le 2$ s, where

$$f(t) = \begin{cases} 0 & t \le 0 \\ t & 0 < t \le 1 \\ e^{-t+1} & 1 < t \le 2 \\ 0 & 2 < t. \end{cases}$$

Four events are defined as: $A = \{f(t) \le 0.75\}$, $B = \{0 \le t \le 1\}$, $C = \{1 < t \le 2\}$, and $D = \{0.5 \le t \le 1.5\}$.

Find: (a) $P(A|B)$, (b) $P(A|C)$, (c) $P(A|D)$, (d) $P(D|A)$.

1.56. Two signals are uniformly sampled every 0.1s from 0 to 4s as follows. A biased coin (where the probability of heads appearing equals 0.35) is tossed to determine the waveform to be sampled: sample $f_1(t)$ if a head appears, otherwise, sample $f_2(t)$. The two signals are

$$f_1(t) = \begin{cases} t & 0 \le t \le 1 \\ 1 & 1 < t \le 3 \\ t - 2 & 3 < t \le 4 \end{cases}$$

and

$$f_2(t) = \begin{cases} 1 & 0 \le t \le 1 \\ -t + 2 & 1 < t \le 2 \\ t - 2 & 2 < t \le 3 \\ 1 & 3 < t \le 4. \end{cases}$$

Evaluate the probability that: (a) the sample is less than or equal to 1/2; (b) the sample is from the interval $0 \le t \le 1$ of $f_1(t)$, given the sample is less than or equal to 1/2.

1.57. Three boxes contain electronic components as listed:

Box 1: 3 capacitors, 3 diodes, 2 resistors;
Box 2: 1 capacitor, 5 diodes;
Box 3: 6 capacitors, 2 diodes, 2 resistors.

A box is chosen at random, then a component is selected at random from the box. (a) Draw a probability tree for the experiment. (b) What is the probability that the component selected is a diode? (c) The component selected was a capacitor. What is the probability that it came from Box 1?

1.58.   The Biomedical Engineering department at FPI has the following numbers of students in each class. Also shown are the percentages of each class which have chosen the Biomechanics track, Bioinstrumentation track, or no track at all.

CLASS	NUMBER OF STUDENTS	OPTION	PERCENT
Freshmen	150	Bioinstrumentation	10%
		Biomechanics	20%
		None	70%
Sophomore	160	Bioinstrumentation	15%
		Biomechanics	15%
		None	70%
Junior	200	Bioinstrumentation	35%
		Biomechanics	45%
		None	20%
Senior	190	Bioinstrumentation	30%
		Biomechanics	30%
		None	40%

A student is chosen at random from the department. (a) Draw a probability tree for this experiment. (b) What is the probability that the student chosen is in the Biomechanics track? (c) Suppose the student chosen was in the Bioinstrumentation track. What is the probability that the student is a junior?

1.59.   A change purse contains three biased coins. A coin is selected at random and tossed. The probability of a head occurring, given coin i is selected is $P(H|C_i) = 0.7, 0.45$, and 0.3, respectively, for $i = 1, 2, 3$. (a) Determine the probability that a head was tossed. (b) Determine the probability that coin three was tossed, given a tail occurred. (c) Suppose a fourth coin is placed in the change purse. After many trials of selecting a coin and then tossing, it is observed that the probability of a head occurring is 0.5. Determine the probability of a head for the fourth coin.

1.60.   One of the passive elements in the circuit shown in Fig. 1.18 is chosen at random. The voltage across the selected element is uniformly sampled every 0.2 s from 0 to 4 s. Determine the probability that: (a) voltage ≥ 1.5, given it is recorded across C_1; (b) voltage ≥ 1.5, given $0 \leq t \leq 0.5$; (c) voltage ≥ 1.5; (d) voltage is recorded across C_1, given that voltage ≥ 1.5.

1.61.   The reliability of some medical diagnosis procedures may not be as good as sometimes indicated. Consider the following problem. A certain test for heart disease is said to

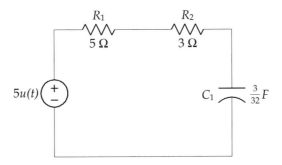

FIGURE 1.18: Circuit for Problem 1.60

be 90% accurate. This can be stated as follows. Let $A = \{$heart disease diagnosed$\}$ and $H = \{$a person has heart disease$\}$. The 90% accuracy is then $P(A|H) = 0.9$ and $P(A^c|H^c) = 0.9$. It is also known by experimental data gathering that $P(H) = 0.01$. Find the probability of a person having heart disease, given that heart disease is diagnosed.

1.62. An experiment involves flipping a fair coin three times. Define the events $A = \{$all heads or all tails$\}$, $B = \{$at least two heads$\}$, and $C = \{$at most two heads result$\}$. Draw a probability tree for this experiment. Are events (a) A and B independent, (b) A and C independent, (c) B and C independent?

1.63. Consider events A and A with $P(A) = 0.45$ and $P(A \cup B) = 0.8$. Determine: (a) the value of $P(B)$ if A and B are independent (b) the value of $P(B)$ if A and B are mutually exclusive (c) if a value of $P(B)$ can be chosen if events A and B are independent and mutually exclusive.

1.64. At least one child in a family having two children is a boy. What is the probability that both children are boys? State your assumptions.

1.65. In the circuit of Fig. 1.19, switches operate independently of one another with each switch having a probability of being closed equal to 0.3. Determine the probability that at any time there is at least (a) one closed path between A and B; (b) one closed path between A and B, given two switches are open.

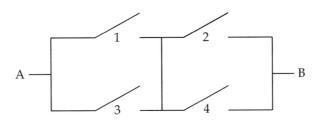

FIGURE 1.19: Circuit for Problem 1.65

1.66. In the circuit of Fig. 1.20, switches operate independently of one another with each switch having a probability of being closed equal to p. Let C, denote the event that switch i is closed, and let C denote the event that there is a closed path from A to B. Find (a) $P(C|C_5)$, (b) $P(C|C_5^c)$, and (c) $P(C)$

1.67. In a ternary digital communication system, the source puts out a symbol (i.e., 0, 1, or 2) every T seconds which is transmitted over a noisy channel to the receiver. The channel introduces errors in which occasionally the symbol received is not the symbol transmitted. Let S_i denote that the symbol i is sent by the source and R_i denote that the receiver observes symbol i. The following probabilities are given:

i	$P(S_i)$	$P(R_0|S_i)$	$P(R_1|S_i)$
0	0.6	0.9	0.05
1	0.3	0.049	0.95
2	0.1	0.1	0.1

1.68. William Smith is a varsity wrestler on his high school team. Without exception, if he does not pin his opponent with his trick move, he loses the match on points. William's trick move also prevents him from ever getting pinned. The probability that William pins his opponent during the first period is 4/10; during the second period is 3/10, given he did not pin his opponent in the first period; and during the third period is 2/10, given he did not pin his opponent in the previous periods. Assume the match is at most three periods. (a) Determine the probability that he pins his opponent during the second period. (b) Determine the probability that he wins the match. (c) Given he won the match, what is the probability he pinned his opponent in the second period. (d) Determine the probability that he wins at least one of his first three matches.

1.69. Doctor Watson has determined that the number of pipes Sherlock Holmes smokes before commencing on a case determines the number of days spent solving that case. Dr. Watson's method is far from certain since Holmes enjoys his pipe enormously.

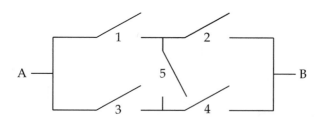

FIGURE 1.20: Circuit for Problem 1.66

However, the number of pipes smoked (never more than three) is always the maximum number of days spent on the case. Being a man of science, Dr. Watson has utilized probability theory to help him describe their cases as follows. With Holmes' reputation, the probability of a three-pipe case is twice as likely as a two- or a one-pipe case. If Holmes smokes N pipes, then the conditional probability he will work D days is given by $P(D|N) = 1/N$, $1 \leq D \leq N$. Now, a successful end to the case is never assured unless Holmes spends the maximum number of days working on the case (i.e., $N = D$ days), otherwise, the conditional probability of success is $2/(2N - D)$. (a) Determine the probability of a successful end to the case. (b) Given Holmes worked fewer days on the case than the number of pipes smoked, determine the probability of a successful end to the case. (c) Holmes completed the case successfully. Determine the probability he smoked two pipes before starting the case. (d) Determine the probability that Holmes works the minimum number of days on a case. (e) Suppose you knew that Holmes was successful. What is the probability of him working the minimum number of days on a case? (f) Should Watson advise Holmes never to stop until he has worked the maximum number of days on a case?

1.70. Consider Problem 1.69 statement. At the conclusion of any case, Holmes believes that he has been successful or else he would not have prematurely stopped the case. It is only after a period of time that he discovers he has been unsuccessful. He then reopens the case and proceeds exactly as before to solve the case. (a) What is the probability that he will reopen a case? (b) What is the probability that he is successful, given he has reopened all unsuccessful cases once? (c) Given that he has reopened a case, determine the probability that it was originally a three pipe case.

1.71. If trouble has a name, it must be baby Leroy. Professor Rensselaer is baby-sitting Leroy for the Smith family and while she is grocery shopping, Leroy disappears. Realizing the gravity of the situation, Ms. Rensselaer assigns these probabilities to determine her course of action during each hour of the search (she does not want any help because she feels awfully foolish losing that child). Leroy is either in the store with a probability of 0.65 or outside the store with a probability of 0.35. The probability she finds him while searching in the store, given Leroy is in the store is 0.3. The probability she finds him while searching outside the store, given Leroy is outside, is 0.45. Assume that Leroy will stay in one location until he is found for all of the following questions. (a) Where should Professor Rensselaer look first to have the best chances of finding Leroy during the first hour of the search? (b) Given Professor Rensselaer looked in the store the first hour and did not find him, what is the probability that Leroy is in the store? (c) Determine the probability that Professor Rensselaer looked outside for the first hour and did not find him and looked outside for the second hour and found him. (d) Suppose there is an equal chance Professor Rensselaer searches in or out of the store. Determine the probability she finds Leroy before the end of the second hour of the search. (e) Suppose there is an equal chance Professor Rensselaer searches in or out of the store and that she finds Leroy during the first hour of the search. What is the probability Leroy remained in the store?

CHAPTER 2

Random Variables

In many applications of probability theory, the experimental outcome space can be chosen to be a set of real numbers; for example, the outcome space for the toss of a single die can be $S = \{1, 2, 3, 4, 5, 6\}$ just as well as the more abstract $S = \{\zeta_1, \zeta_2, \ldots, \zeta_6\}$, where ζ_i represents the outcome that i dots appear on the top face of the die. In virtually all applications, a suitable mapping can be found from the abstract outcome space to the set of real numbers. Once this mapping is performed, all computations and analyses can be applied to the resulting real numbers instead of to the original abstract outcome space. This mapping is called a *random variable*, and enables us to develop a uniform collection of analytical tools which can be applied to any specific problem. Furthermore, this mapping enables us to deal with real numbers instead of abstract entities.

2.1   MAPPING

A random variable $x(\cdot)$ is a mapping from the outcome space S to the extended real number line: $-\infty \leq x(\zeta) \leq +\infty$, for all $\zeta \in S$. This mapping is illustrated in Fig. 2.1. We will often express such a mapping as $x : S \mapsto R^*$, where $R^* = R \cup \{+\infty, -\infty\}$ is the set of extended real numbers. The notation $f : A \mapsto B$ can be read as "f is a function mapping elements in A to elements in B." The reader is undoubtedly familiar with real valued functions of real variables, e.g., $\sin : R \mapsto [-1, 1]$. The mapping performed by a random variable (RV) is a bit different, in that the domain S is (in general) a set of abstract elements. It is also important to note the distinction between a probability measure function (with argument a *set*) and a random variable which has an *element of a set* as an argument.

For many experiments, such as the measurement of a voltage or current, the observed phenomenon is inherently a real number; in others, such as drawing a card or an item from inventory, the observed entity is abstract. From another perspective, in some instances the association of a number to an abstract experimental outcome is more natural than in others. One can, for example, number each of the cards in a deck. The mapping performed by a random variable enables us to apply the mathematics of real numbers to aid in problem solving. In many applications, the problem solver can choose a mapping that simplifies the problem

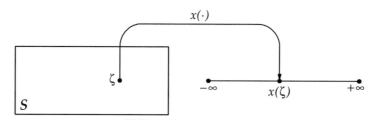

FIGURE 2.1: A random variable $x(\cdot)$ maps each outcome in S to an extended real number

solution. Recall that with virtually any application of probability theory, probabilities need to be computed for events comprised of elements of the experimental outcome space. The mapping may be one-to-one, so that to each value of the RV x there corresponds one and only one $\zeta \in S$, and thus knowledge of the value of the random variable determines uniquely the experimental outcome ζ. In many cases, the mapping performed by the RV x is a many-to-one mapping: for each value of the RV there may correspond many experimental outcomes. In addition, more than one RV may be defined on the same outcome space.

An arbitrary mapping from the outcome space S to the set of extended real numbers R^* is not necessarily a bona fide RV—some additional restrictions are needed. In this section, we focus on properties of a general mapping from S to R^*.

Example 2.1.1. *Consider the card–drawing experiment. Define the mapping*

$$x(\zeta) = 1, 2, \ldots, 13,$$

if the card drawn is $\zeta = 2, 3, 4, \ldots, 10,$ Jack, Queen, King, Ace, respectively. Define the mapping $y(\zeta) = 1, 2, 3, 4$ if the suit of the card drawn is Clubs, Hearts, Spades, Diamonds, respectively. Note that ζ, the experimental outcome, is the card drawn so that if $\zeta =$ King of Hearts then $x(\zeta) = 12$, and $y(\zeta) = 2$.

(a) *Define the mapping $z(\zeta) = x(\zeta) + (y(\zeta) - 1) \times 13$. Determine the possible outcomes of the card drawing if $z(\zeta)$ is known to lie in the interval $[25, 28]$.*

(b) *Define the mapping $w(\zeta) = x(\zeta) + y(\zeta)$, and suppose that the value of w is 7. Determine the possible outcomes that occurred.*

Solution

(a) Since $x \in \{1, 2, \ldots, 13\}$, we know that if $y(\zeta) = 1$, then $z(\zeta) \in \{1, 2, \ldots, 13\}$; if $y(\zeta) = 2$, then $z(\zeta) \in \{14, 15, \ldots, 26\}$, etc. Since we know that $25 \leq z < 28$ and z is an integer, we know that $z(\zeta) \in \{25, 26, 27\}$. Consequently, either $y = 2$ and $x \in \{12, 13\}$, or $y = 3$ and $x = 1$. Thus, the possible outcomes are

$$\zeta \in \{\text{King of Hearts, Ace of Hearts, 2 of Spades}\}.$$

Note that the mapping performed by z is a one-to-one mapping.

(b) Since y can only take on four possible values, we simply consider all possibilities. If $y = 1$, then $x = w - y = 7 - 1 = 6$; if $y = 2$, then $x = 5$; if $y = 3$, then $x = 4$; if $y = 4$, then $x = 3$. Hence, the possible outcomes are

$$\zeta \in \{7 \text{ of Clubs}, \ 6 \text{ of Hearts}, \ 5 \text{ of Spades}, \ 4 \text{ of Diamonds}\}.$$

This is an example of a many-to-one mapping. ∎

Events on a probability space $(S, \mathcal{F}, P)$ are sets of outcomes in S. We are concerned with the images of these events as a result of the mapping performed by a RV x. We use the notation

$$x(A) = \{x(\zeta) : \zeta \in A\} \tag{2.1}$$

to denote the image of the event $A \in \mathcal{F}$. Similarly, the inverse image of a set of real numbers B is denoted by

$$x^{-1}(B) = \{\zeta \in S : x(\zeta) \in B\}. \tag{2.2}$$

It is of interest to note that the set function $x^{-1}(B)$ defined above always exists, whereas the point function $x^{-1}(b)$ may fail to exist for some real values of b.

Example 2.1.2. *A fruit bowl contains one gallon of cold water, 10 apples, 6 oranges, and 15 bananas. Baby Leroy (our experimenter for today) dips his hand in the bowl and extracts one item of fruit and a quantity of water. The RV x is defined as follows:*

$$x(\zeta) = i(\zeta) + w(\zeta),$$

where $i(\zeta) = 1, 2,$ or 3, if the item of fruit is an apple, orange, or banana, respectively, and $w(\zeta)$ is the amount of water in ounces. Assume that baby Leroy's hand, together with the fruit, holds no more than 1.5 ounces of water. For each of the following sets, find $A = x^{-1}(B)$:

(a) $B = \{1\}$

(b) $B = \{1.2, 3.2\}$

(c) $B = [2.1, 2.9) \cup \{1.1\}$.

Solution

(a) $A = \{\text{an apple with } no \text{ water}\}$

(b) $A = \{\text{an apple with 0.2 Oz. water, an orange with 1.2 Oz. water, a banana with 0.2 Oz. water}\}$

(c) $A = \{$an apple with 1.1–1.5 Oz. water, an apple with 0.1 Oz. water, an orange with 0.1–0.9 Oz. water$\}$ ∎

Example 2.1.3.   *Let $x(t) = \sin(t)$. Find (a) $x^{-1}(\{1\})$, and (b) $x^{-1}(\{2\})$.*

Solution

(a)   We recall that $\sin\left(\frac{\pi}{2}\right) = 1$ and that $\sin(t)$ is periodic in t with period 2π. Consequently,

$$x^{-1}(\{1\}) = \left\{t = k2\pi + \frac{\pi}{2} : k = 0, \pm1, \pm2, \ldots\right\}.$$

(b)   Since $-1 \leq \sin(t) \leq 1$ for all real values of t, we conclude that $x^{-1}(2)$ does not exist, so that $x^{-1}(\{2\}) = \varnothing$.   ∎

The following theorems establish properties of a general mapping from elements in S to R^*. The mapping $g : S \mapsto R^*$ is considered to be defined as a point function. Properties considered below concern the images and inverse images of sets under the mapping performed by g. Because of our interest in random variables, we take the range space to be R^*; however, the properties remain true with an arbitrary range space. On the first reading, the remainder of this section may be skipped—it is used only for the proof of Theorem 2.1.1.

Theorem 2.1.1.   *Let $A_1, A_2, \ldots$ be subsets of S and let $g : S \mapsto R^*$. Then*

$$g\left(\bigcup_i A_i\right) = \bigcup_i g(A_i) \tag{2.3}$$

and

$$g\left(\bigcap_i A_i\right) \subset \bigcap_i g(A_i). \tag{2.4}$$

Proof.   Let

$$g_1 \in g\left(\bigcup_i A_i\right)$$

Then there exists a $\zeta_1 \in A_{i_1}$ for at least one i_1 such that $g(\zeta_1) = g_1$; hence

$$g_1 \in \bigcup_i g(A_i),$$

so that

$$g\left(\bigcup_i A_i\right) \subset \bigcup_i g(A_i).$$

Now let

$$g_1 \in \bigcup_i g(A_i).$$

Then $g_1 \in g(A_{i_1})$ for at least one i_1, so that there exists a $\zeta_1 \in A_{i_1}$ with $g(\zeta_1) = g_1$; hence

$$g_1 \in g\left(\bigcup_i A_i\right),$$

so that

$$\bigcup_i g(A_i) \subset g\left(\bigcup_i A_i\right),$$

and (2.3) is satisfied.

Let

$$g_1 \in g\left(\bigcap_i A_i\right).$$

Then there exists a $\zeta_1 \in A_i$ with $g_1 = g(\zeta_1) \in g(A_i)$ for every i, yielding (2.4). ∎

Example 2.1.4. *Give an example where (2.4) is not an equality.*

Solution. Let $S = R^*$, $g(\zeta) = u(\zeta)$, $A_1 = [1, 2]$, and $A_2 = [3, 4]$. Then $g(A_1 \cap A_2) = g(\varnothing) = \varnothing$ and $g(A_1) \cap g(A_2) = \{1\}$. ∎

Theorem 2.1.2. *Let $A \subset S$, $B \subset R^*$, $B_i \subset R^*$, $i = 1, 2, \ldots$, and let $g : S \mapsto R^*$. Then*

$$g^{-1}(B^c) = (g^{-1}(B))^c, \tag{2.5}$$

$$g^{-1}\left(\bigcup_i B_i\right) = \bigcup_i g^{-1}(B_i), \tag{2.6}$$

$$g^{-1}\left(\bigcap_i B_i\right) = \bigcap_i g^{-1}(B_i), \tag{2.7}$$

and

$$g(g^{-1}(B) \cap A) = B \cap g(A). \tag{2.8}$$

If $B \subset g(S)$ then

$$g(g^{-1}(B)) = B. \tag{2.9}$$

If $A \subset S$ then

$$A \subset g^{-1}(g(A)). \tag{2.10}$$

Proof.   We have

$$g^{-1}(B^c) = \{\zeta : g(\zeta) \notin B\} = \{\zeta : g(\zeta) \in B\}^c = (g^{-1}(B))^c,$$

yielding (2.5).

Equation (2.6) follows from

$$\left\{\zeta : g(\zeta) \in \bigcup_i B_i\right\} = \bigcup_i \{\zeta : g(\zeta) \in B_i\}.$$

Let

$$B = \bigcup_i B_i^c.$$

Then

$$B^c = \bigcap_i B_i.$$

Using (2.5) and (2.6) we obtain

$$g^{-1}(B^c) = (g^{-1}(B))^c = \left(\bigcup_i g^{-1}(B_i^c)\right)^c = \bigcap_i g^{-1}(B_i),$$

so that (2.7) is satisfied.

We obtain

$$\begin{aligned}
g(g^{-1}(B) \cap A) &= \{g(\zeta) : \zeta \in g^{-1}(B) \cap A\} \\
&= \{g(\zeta) : g(\zeta) \in B, \zeta \in A\} \\
&= B \cap g(A),
\end{aligned}$$

yielding (2.8).

Letting $A = S$ and $B \subset g(S)$ in (2.8) yields (2.9).

Let $A \subset S$. By definition

$$g^{-1}(g(A)) = \{\zeta : g(\zeta) \in g(A)\}.$$

Clearly, if $\zeta_1 \in A$ then $g(\zeta_1) \in g(A)$ so that $\zeta_1 \in g^{-1}(g(A))$, and (2.10) is satisfied. ∎

Example 2.1.5.   *Find an example for which (2.10) is not an equality.*

Solution.   Let $S = R^*$, $g(\zeta) = u(\zeta)$, and $A = [0, 2]$. Then $g(A) = \{1\}$ and $g^{-1}(g(A)) = [0, \infty]$. ∎

Theorem 2.1.3. *Let $f : S \mapsto U$, $g : U \mapsto R^*$, and $h : S \mapsto R^*$, with $h(\zeta) = g(f(\zeta))$. Then*

$$h^{-1}(B) = f^{-1}(g^{-1}(B)) \tag{2.11}$$

for any $B \subset R^*$.

Proof. By definition

$$\begin{aligned} h^{-1}(B) &= \{\zeta : g(f(\zeta)) \in B\} \\ &= \{\zeta : f(\zeta) \in g^{-1}(B)\} \\ &= f^{-1}(g^{-1}(B)). \end{aligned}$$ ■

Theorem 2.1.4. *Let $g : S \mapsto R^*$, let $\mathcal{B}$ be a nonempty collection of subsets of R^* and let $\sigma(\mathcal{B})$ denote the minimal sigma field containing $\mathcal{B}$. Then*

$$\sigma(g^{-1}(\mathcal{B})) = g^{-1}(\sigma(\mathcal{B})). \tag{2.12}$$

Proof. Since $g^{-1}(R^*) = S$, (2.5) and (2.6) show that $g^{-1}(\sigma(\mathcal{B}))$ is a σ-field of subsets of S. Since $g^{-1}(\mathcal{B}) \subset g^{-1}(\sigma(\mathcal{B}))$, we then obtain

$$\sigma(g^{-1}(\mathcal{B})) \subset g^{-1}(\sigma(\mathcal{B})).$$

Consider the collection of subsets of R^* specified by

$$\varepsilon = \{\varepsilon : g^{-1}(\varepsilon) \in \sigma(g^{-1}(\mathcal{B}))\}.$$

Since $g^{-1}(R^*) = S$, (2.5) and (2.6) reveal that ε is a σ-field of subsets of R^*. Now, since $\mathcal{B} \subset \varepsilon$ we find $\sigma(\mathcal{B}) \subset \varepsilon$ so that

$$g^{-1}(\sigma(\mathcal{B})) \subset g^{-1}(\varepsilon) \subset \sigma(g^{-1}(\mathcal{B})).$$ ■

Drill Problem 2.1.1. *Consider the mapping $g : R^* \mapsto [-1, 1]$, with $g(\zeta) = \sin(2\zeta)$. Find: (a) $g^{-1}(\{0\})$, (b) $g^{-1}(\{-1\})$, and (c) $g^{-1}([0, 1])$.*

Answers: $\bigcup_{k=-\infty}^{\infty}[2k\frac{\pi}{2}, (2k + 1)\frac{\pi}{2}]$,

$$\left\{\left(k + \frac{3}{4}\right)\pi : k = 0, \pm 1, \pm 2, \ldots\right\}, \quad \left\{k\frac{\pi}{2} : k = 0, \pm 1, \pm 2, \ldots\right\}.$$

Drill Problem 2.1.2. *Let $g : R^* \mapsto [0, 1]$, with $g(\zeta) = \zeta(u(\zeta) - u(\zeta - 1))$. Find $g^{-1}((0, 1))$ and $g^{-1}([0, 1))$.*

Answers: $(0, 1)$, R^*.

2.2   MEASURABLE FUNCTIONS

Let $(S, \mathcal{F}, P)$ be a probability space, and let $x : S \mapsto R^*$. By definition, we may compute the probability that event A occurs using the probability measure P. In the following sections, we examine techniques for computing the probability that x takes on a value in a Borel set B. This probability, denoted as $\text{Prob}(x(\zeta) \in B)$, is a legitimate event probability only if there is an event $A \in \mathcal{F}$ such that the image of A under the mapping x is the Borel set B; i.e., only if $B = \{x(\zeta) : \zeta \in A\}$ for some $A \in \mathcal{F}$. If such an event A exists, we then have $P(A) = \text{Prob}(x(\zeta) \in B)$. In order that the mapping $x : S \mapsto R^*$ be a bona fide random variable, we require that such an event $A \in \mathcal{F}$ exist for every Borel set B. Such a mapping is known as a *measurable function*.

Definition 2.2.1. *Let $(S, \mathcal{F})$ be a measurable space and let $x : S \mapsto R^*$. If $x^{-1}(B) \in \mathcal{F}$ for each Borel set B of R^*, then we say that x is $\mathcal{F}$-measurable, or simply* **measurable**.

Definition 2.2.2. *A real* **random variable** *(RV) on the probability space $(S, \mathcal{F}, P)$ is an R^*-valued measurable function with domain S. A RV is allowed to take on the values $\pm\infty$, but only with probability zero.*

Theorem 2.2.1. *Let $(S, \mathcal{F})$ be a measurable space, and let $g : S \mapsto R^*$. Then g is measurable iff*

$$g^{-1}([-\infty, \alpha]) \in \mathcal{F}, \quad \text{for all } \alpha \in R^*. \tag{2.13}$$

Proof. Since $[-\infty, \alpha]$ is a Borel set for each value of α, if g is measurable then (2.13) is satisfied.

Let $\varepsilon = \{[-\infty, \alpha] : \alpha \in R^*\}$ and assume (2.13) is satisfied. Then $g^{-1}(\varepsilon) \subset \mathcal{F}$. Let $\mathcal{B}$ denote the collection of Borel sets of R^*, and note that

$$(a, b) = [-\infty, a]^c \cap \bigcup_{n=1}^{\infty} \left[-\infty, b - \frac{1}{n}\right] \in \sigma(\varepsilon).$$

Hence $\mathcal{B} = \sigma(\varepsilon)$. Applying Theorem 2.1.4, we have

$$g^{-1}(\mathcal{B}) = g^{-1}(\sigma(\varepsilon)) = \sigma(g^{-1}(\varepsilon)) \subset \mathcal{F},$$

so that g is measurable. ∎

Example 2.2.1. *A fair die is tossed once. Let $S = \{1, 2, 3, 4, 5, 6\}$, $A = \{1, 2, 3\}$, and $\mathcal{F} = \{\varnothing, S, A, A^c\}$.*

(a) *Let $x : S \mapsto R^*$ be defined by $x(\zeta) = \zeta$. Is x a RV?*

(b) *Let $y : S \mapsto R^*$ be defined by*

$$y(\zeta) = \begin{cases} 5, & \zeta \in A \\ 10, & \zeta \in A^c. \end{cases}$$

Is y a RV?

Solution

(a) To apply Theorem 2.2.1, we try to find an interval with inverse image not belonging to $\mathcal{F}$. Since $x^{-1}([-\infty, 2]) = \{1, 2\} \notin \mathcal{F}$, x is not a RV. Note that redefining $\mathcal{F}$ to be the collection of all subsets of S would make x a bona fide RV.

(b) We have

$$y^{-1}([-\infty, \alpha]) = \begin{cases} \varnothing, & \alpha < 5 \\ A, & 5 \le \alpha < 10 \\ S, & 10 \le \alpha. \end{cases}$$

Hence, y is a RV. ∎

The example above suggests that a nonmeasurable function can often be made measurable by considering a different σ-field. In the sequel, we assume the RVs considered are measurable functions on the probability space $(S, \mathcal{F}, P)$.

Example 2.2.2. *A box contains a collection of resistors, inductors, capacitors, and transistors. One component is drawn from the box, with*

$$P(\{\text{resistor}\}) = 0.1,$$

$$P(\{\text{inductor}\}) = 0.3,$$

$$P(\{\text{capacitor}\}) = 0.2,$$

and

$$P(\{\text{transistor}\}) = 0.4.$$

Define the random variable x by $x(\zeta) = 5, 0, -1.5,$ and 2, respectively, for

$$\zeta = \text{resistor, inductor, capacitor, and transistor.}$$

(a) *Determine the function*

$$F_x(\alpha) = P(\{\zeta : x(\zeta) \le \alpha\})$$

for all real α. The function $F_x(\alpha)$ is called the cumulative distribution function for the random variable x, and will prove to be very useful throughout our remaining work in probability theory.

(b) *Use $F_x(\alpha)$ to find: (i) $P(\{\text{resistor}\})$, (ii) $P(\{\text{inductor, capacitor, transistor}\})$, and (iii) $P(\{\text{inductor, transistor}\})$.*

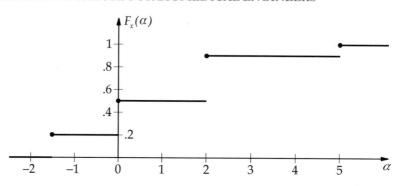

FIGURE 2.2: Cumulative distribution function for Example 2.2.2

Solution

(a)   From the given information, we easily find

$$F_x(\alpha) = \begin{cases} 0, & \text{if } \alpha < -1.5 \\ 0.2, & \text{if } -1.5 \le \alpha < 0 \\ 0.5, & \text{if } 0 \le \alpha < 2 \\ 0.9, & \text{if } 2 \le \alpha < 5 \\ 1, & \text{if } 5 \le \alpha. \end{cases}$$

The function $F_x(\alpha)$ is illustrated in Fig. 2.2.

(b)

(i) Defining $F_x(\alpha^-)$ to denote the value of F_x just to the left of α, we have

$$P(\{\text{resistor}\}) = P(\{\zeta : x(\zeta) = 5\}) = F_x(5) - F_x(5^-) = 0.1.$$

(ii) Using the correspondence between the values of the random variable x and the selected components, we obtain

$$\begin{aligned} P(\{\text{inductor, capacitor, transistor}\}) &= P(\{\zeta : x(\zeta) \in \{0, -1.5, 2\}\}) \\ &= P(\{\zeta : -1.5 \le x(\zeta) \le 2\}) \\ &= F_x(2) - F_x(-1.5^-) \\ &= 0.9. \end{aligned}$$

(iii) We have

$$\begin{aligned} P(\{\text{inductor, transistor}\}) &= P(\{\zeta : x(\zeta) \in \{0, 2\}\}) \\ &= F_x(0) - F_x(0^-) + F_x(2) - F_x(2^-) \\ &= 0.5 - 0.2 + 0.9 - 0.5 = 0.7. \end{aligned}$$ ∎

The example above introduces the important concept of a cumulative distribution function and how it can be used to compute event probabilities. This concept is expanded in the following section.

Drill Problem 2.2.1. *An urn contains seven red marbles and three white marbles. Two marbles are drawn from the urn one after the other without replacement. Let the random variables x and y denote the total number of red and (respectively) white marbles selected. Find: (a) $P(\zeta : x(\zeta) = 0)$, (b) $P(\zeta : x(\zeta) = y(\zeta))$, (c) $x^{-1}([0.5, 5))$, and (d) the smallest sigma-field that can be used so that $z(\zeta) = x(\zeta) + y(\zeta)$ is a random variable.*

Answers: $\{\varnothing, S\}$, 1/15, $\{R_1 R_2, R_1 W_2, W_1 R_2\}$, 7/15.

Drill Problem 2.2.2. *Consider the experiment of tossing a fair tetrahedral die (with faces labeled 0, 1, 2, 3) twice. Let x be a random variable denoting the sum of the numbers tossed. Determine the probability that x takes the values (a) 0, (b) 2, (c) 3, and (d) 5.*

Answers: 2/16, 1/16, 4/16, 3/16.

2.3   CUMULATIVE DISTRIBUTION FUNCTION

By definition, the probability that a RV x (defined on the probability space $(S, \mathcal{F}, P)$) takes on a value in any particular Borel set B can be determined from $P(x^{-1}(B))$. In this section, we develop the concept of a cumulative distribution function (CDF) for the RV x which enables us to compute the desired probabilities directly without explicitly making use of the probability measure P.

Definition 2.3.1. *Let x be a RV on the probability space $(S, \mathcal{F}, P)$. Define $F_x : R^* \mapsto [0, 1]$ by*

$$F_x(\alpha) = P(x^{-1}([-\infty, \alpha])) = P(\{\zeta : -\infty \leq x(\zeta) \leq \alpha\}).$$

*The function F_x is the **cumulative distribution function** (CDF) for the RV x.*

Note that the RV x and the probability measure P determine the CDF F_x. Furthermore,

$$x^{-1}([-\infty, \alpha]) = x^{-1}(\{-\infty\}) \cup x^{-1}((-\infty, \alpha]),$$

and that $P(x^{-1}(\{-\infty\})) = 0$, so that

$$F_x(\alpha) = P(x^{-1}([-\infty, \alpha])) = P(\{\zeta : -\infty < x(\zeta) \leq \alpha\}).$$

Using the relative frequency approach to probability assignment, a CDF can be estimated as follows. Suppose that a RV x takes on the values $x_1, x_2, \ldots, x_n$ in n trials of an experiment. The function

$$\hat{F}_x(\alpha) = \frac{1}{n} \sum_{i=1}^{n} u(a - x_i) \qquad (2.14)$$

is an estimate of the CDF $F_x(\alpha)$, where $u(\cdot)$ is the unit step function. This estimate $\hat{F}_x(\alpha)$ will be referred to as the **empirical distribution function** for the RV x. Let n_α denote the number of times the RV x is observed to be less than or equal to α in n trials of the experiment. Note that $\hat{F}_x(\alpha) = \frac{n_\alpha}{n}$.

The empirical distribution can be applied to the "random sampling" of a waveform $w(\cdot)$. Let $S = \{0, 1, \ldots, N-1\}$, and $P(\zeta = i) = 1/N$ for $i \in S$. Let the RV $x(\zeta) = w(a + \zeta T)$ for each $\zeta \in S$, where $T > 0$ is the sampling period. The CDF for x is

$$F_x(\alpha) = \frac{1}{N} \sum_{k=0}^{N-1} u(\alpha - w(a + kT)).$$

Definition 2.3.2. *The function $f : R^* \mapsto R$ is **right-continuous** at x if $f(x^+) = f(x)$, where*

$$f(x^+) = \lim_{h \to 0} f(x + h), \qquad (2.15)$$

*and **left-continuous** at x if $f(x^-) = f(x)$, where*

$$f(x^-) = \lim_{h \to 0} f(x - h) = f(x). \qquad (2.16)$$

*The limits are taken through positive values of h. We say that f is **continuous at** x if $f(x^+) = f(x^-) = f(x)$ and simply **continuous** if f is continuous for all real x.*

Theorem 2.3.1 (Properties of CDF). *Let x be a RV on the probability space $(S, \mathcal{F}, P)$, and let F_x be the CDF for x. Then*

(i) *$F_x(a) \le F_x(b)$ for all $a < b$; i.e., F_x is monotone nondecreasing;*
(ii) *F_x is right-continuous;*
(iii) *$F_x(-\infty) = 0$;*
(iv) *$F_x(\infty) = 1$;*
(v) *$P(x^{-1}((a, b])) = F_x(b) - F_x(a)$ for all $a < b$;*
and
(vi) *$P(x^{-1}(\{a\})) = F_x(a) - F_x(a^-)$.*

Proof

(i) For all $a < b$,
$$\begin{aligned}
F_x(b) &= P(x^{-1}((-\infty, b])) \\
&= P(x^{-1}((-\infty, a]) \cup x^{-1}((a, b])) \\
&= P(x^{-1}((-\infty, a])) + P(x^{-1}((a, b])) \\
&= F_x(a) + P(x^{-1}((a, b])) \\
&\ge F_x(a).
\end{aligned}$$

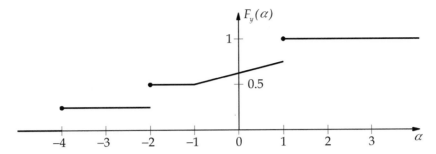

FIGURE 2.3: CDF for Example 2.3.1

(ii) Let $a = \alpha$, $b = \alpha + h$ and $h > 0$. From (i) above,

$$F_x(\alpha + h) = F_x(\alpha) + P(x^{-1}((\alpha, \alpha + h])).$$

Since $(\alpha, \alpha + h] \to \varnothing$ as $h \to 0$, we have $x^{-1}((\alpha, \alpha + h]) \to \varnothing$ and hence $P(x^{-1}((\alpha, \alpha + h])) \to 0$ as $h \to 0$.

(iii) and (iv) follow from the definition of a random variable, requiring that $P(x^{-1}(\{\pm\infty\})) = 0$.

(v) From (i) above, $P(x^{-1}((a, b])) = F_x(b) - F_x(a)$ for all $a < b$.

(vi) follows from (v) replacing a with a^- and b with a. ■

Applying the above theorem, the probability that a RV x takes on a value in an arbitrary Borel set $\mathcal{B}$ can be determined directly from the CDF F_x. Consequently, the CDF determines a probability measure P_F on the measurable space $(R^*, \mathcal{B})$, where $\mathcal{B}$ is the Borel field for R^*. If one is only interested in the RV x, then one need only consider the probability space $(R^*, \mathcal{B}, P_F)$. Any function F_x mapping R^* to R and satisfying properties (i)–(iv) of Theorem 2.3.1 is a valid CDF for determining P_F on the probability space $(R^*, \mathcal{B}, P_F)$.

From now on, we will often shorten the notation for probabilities. For example, the expressions $P(\{\zeta : \zeta \in x^{-1}((a, b])\})$, $P(x^{-1}((a, b]))$, $P(a < x(\zeta) \le b)$, and $P(a < x \le b)$ are all equivalent.

Example 2.3.1.   *The RV y has CDF F_y shown in Fig. 2.3. Find (a) $P(y = -2)$, (b) $P(-2 \le y < -1.5)$, and (c) $P(-0.5 < y < 1)$.*

Solution

(a) Since $P(y = -2) = P(-2^- < y \le 2)$, we find

$$P(y = -2) = F_y(-2) - F_y(-2^-) = 0.5 - 0.25 = 0.25.$$

(b) Since $P(-2 \le y < -1.5) = P(-2^- < y \le -1.5^-)$, we find

$$P(-2 \le y < -1.5) = F_y(-1.5^-) - F_y(-2^-) = 0.5 - 0.25 = 0.25.$$

(c)   Since $P(-0.5 < y < 1) = P(-0.5 < y \le 1^-)$, we obtain

$$P(-0.5 < y < 1) = F_y(1^-) - F_y(-0.5) = 0.75 - \left(0.5 + \frac{1}{4}(0.75 - 0.5)\right) = 3/16. \quad \blacksquare$$

There are two basic categories of RVs with which we will be concerned: discrete RVs and continuous RVs. The RV y in the above example is a mixed RV—a RV with both a discrete part and a continuous part. The discrete part corresponds to the jumps in the CDF and the continuous part corresponds to the interval $(-1, 1)$ where the CDF is increasing in a continuous manner. We now define these types of RVs. Note that the CDF has a **jump discontinuity** at α if $F_x(\alpha) - F_x(\alpha^-) \ne 0$. Furthermore, since a CDF is right-continuous and bounded, the only kind of discontinuity a CDF may have is a jump discontinuity. Similarly, the number of discontinuities a CDF may have is countable.

2.3.1   Discrete Random Variables

Discrete random variables take on at most a countable number of values. The resulting CDF is a jump function. Probabilities for discrete random variables are often easily found with the aid of the probability mass function—which can be found from the CDF.

Definition 2.3.3.   *A RV x on $(S, \mathcal{F}, P)$ is a **discrete RV** if the CDF F_x is a jump function; i.e., iff there exists a countable set $D_x \subset R$ such that*

$$P(\{\zeta : x(\zeta) \in D_x\}) = 1. \tag{2.17}$$

The function

$$p_x(\alpha) = P(\{\zeta : x(\zeta) = \alpha\}) = F_x(\alpha) - F_x(\alpha^-) \tag{2.18}$$

*is called the **probability mass function** (PMF) for the discrete RV x. The set of points D_x for which the PMF is nonzero is called the **support set** for p_x.*

Theorem 2.3.2.   *Let x be a discrete RV on the probability space $(S, \mathcal{F}, P)$. Then*

$$F_x(\alpha) = \sum_{\alpha' \in D_x \cap (-\infty, \alpha]} P_x(a'), \tag{2.19}$$

$p_x(\alpha) \ge 0$ for all real α,

$$\sum_\alpha P_x(\alpha) = 1, \tag{2.20}$$

and

$$P(\{\zeta : x(\zeta) \in A\}) = \sum_{\alpha \in A} P_x(a). \tag{2.21}$$

All summation indices are assumed to belong to the support set D_x.

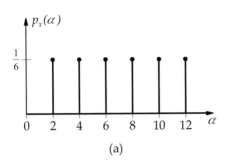

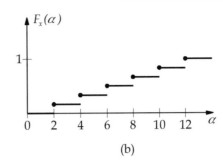

(a) (b)

FIGURE 2.4: (a) PMF and (b) CDF for Example 2.3.2

Proof. The proof is a straightforward application of the definitions of PMF and CDF. ∎

Any function p_x mapping R^* to R which has support set D_x and satisfies

$$p_x(\alpha) \geq 0 \quad \text{for all real } \alpha, \tag{2.22}$$
$$p_x(-\infty) = p_x(+\infty) = 0, \tag{2.23}$$

and

$$\sum_{\alpha \in D_x} P_x(\alpha) = 1 \tag{2.24}$$

is a valid PMF.

Example 2.3.2. *A fair die is tossed once. The RV x is twice the number of dots appearing on the die face. Find: (a) the PMF p_x, (b) the CDF F_x, (c) $P(3.5 < x \leq 8)$.*

Solution

(a) The support set for p_x is $D_x = \{2, 4, 6, 8, 10, 12\}$. The PMF for x is

$$p_x(\alpha) = \begin{cases} \dfrac{1}{6}, & \alpha \in D_x \\ 0, & \text{otherwise,} \end{cases}$$

and is shown in Fig. 2.4a.

(b) The CDF is

$$F_x(\alpha) = \sum_{i=1}^{6} \frac{1}{6} u(\alpha - 2i).$$

which can be expressed as

$$F_x(\alpha) = \begin{cases} 0, & \alpha < 2 \\ \dfrac{k}{6}, & 2k \leq \alpha < 2k + 2, \ k = 1, 2, 3, 4, 5 \\ 1, & \alpha \geq 12. \end{cases}$$

The CDF F_x is shown in Fig. 2.4b.

(c)   We have

$$P(3.5 < x \leq 8) = \sum_{\alpha \in \{4,6,8\}} P_x(\alpha) = \frac{3}{6} = \frac{1}{2}.$$ ■

The set of points D_x for which a discrete RV has nonzero probability (the support set) may contain an infinite number of elements. Whenever there are many points in D_x, it is highly desirable to express the CDF in "closed form." In many cases, the discrete RVs of interest are lattice RVs.

Definition 2.3.4.   *The RV x is a **lattice random variable** if there exist real constants a and h such that h > 0 and*

$$\sum_{k=-\infty}^{\infty} P(x = kh + a) = 1. \tag{2.25}$$

*The value of h is called a **span** of the lattice RV.*

Two of the most basic integrals we encounter are

$$\int_0^a x^k \, dx = \frac{a^{k+1}}{k+1}, \quad k \geq 0,$$

and

$$\int_a^b e^{\alpha x} \, dx = \frac{e^{\alpha b} - e^{\alpha a}}{\alpha}.$$

The following lemmas provide corresponding basic results for summations which arise with lattice RV probability calculations.

Lemma 2.3.1.   *Define*

$$\gamma_k^{[\ell]} = \prod_{i=0}^{l-1}(k - i) = \begin{cases} \overbrace{k(k-1)\cdots(k-\ell+1)}^{\ell \text{ terms}}, & \ell = 1, 2, \ldots \\ 1, & \ell \leq 0. \end{cases} \tag{2.26}$$

Then for integer n ≥ 0

$$\sum_{k=0}^{n} \gamma_k^{[\ell]} = \frac{1}{\ell+1} \gamma_n^{[\ell+1]}, \quad \ell = 0, 1, \ldots \tag{2.27}$$

For integer $n \leq 0$

$$\sum_{k=n}^{0} \gamma_k^{[\ell]} = -\frac{1}{\ell+1}\gamma_n^{[\ell+1]}, \quad \ell = 0, 1, \ldots \tag{2.28}$$

Proof. First consider $n \geq 0$. Note that $\gamma_k^{[\ell]} = 0$ for $k = 0, 1, \ldots, \ell - 1$. For $k = \ell, \ell + 1, \ldots, n$ we have

$$\gamma_{k+1}^{[\ell+1]} - \gamma_k^{[\ell+1]} = \prod_{i=0}^{\ell}(k+1-i) - \prod_{i=0}^{\ell}(k-i)$$

$$= (k+1)\prod_{i=1}^{\ell}(k-(i-1)) - (k-\ell)\gamma_k^{[\ell]}$$

$$= (k+1-k+\ell)\gamma_k^{[\ell]} = (\ell+1)\gamma_k^{[\ell]}.$$

Summing from $k = \ell$ to $k = n$ we obtain

$$(\ell+1)\sum_{k=\ell}^{n}\gamma_k^{[\ell]} = \sum_{k=\ell}^{n}\left(\gamma_{k+1}^{[\ell+1]} - \gamma_k^{[\ell+1]}\right) = \gamma_{n+1}^{[\ell+1]} - \gamma_\ell^{[\ell+1]} = \gamma_{n+1}^{[\ell+1]},$$

from which the desired result (2.27) follows.

Now consider $n \leq 0$. For integer $k \leq -1$

$$\gamma_k^{[\ell+1]} - \gamma_{k+1}^{[\ell+1]} = (k-\ell-k-1)\gamma_k^{[\ell]} = -(\ell+1)\gamma_k^{[\ell]}.$$

Summing from $k = n$ to $k = 0$ we obtain

$$-(\ell+1)\sum_{k=n}^{0}\gamma_k^{[\ell]} = \sum_{k=n}^{0}\gamma_k^{[\ell+1]} - \sum_{m=n+1}^{0}\gamma_k^{[\ell+1]} = \gamma_n^{[\ell+1]},$$

where we have used the fact that $\gamma_0^{[\ell+1]} = 0$. This establishes (2.28). ∎

In particular, the above lemma yields

$$\sum_{k=0}^{n} 1 = \sum_{k=0}^{n}\gamma_k^{[0]} = \gamma_{n+1}^{[1]} = n+1, \tag{2.29}$$

$$\sum_{k=0}^{n} k = \sum_{k=0}^{n}\gamma_k^{[1]} = \frac{\gamma_{n+1}^{[2]}}{2} = \frac{(n+1)n}{2}, \tag{2.30}$$

and

$$\sum_{k=0}^{n} k^2 = \sum_{k=0}^{n}k(k+1) + \sum_{k=0}^{n}k = \frac{\gamma_{n+1}^{[3]}}{3} + \frac{(n+1)n}{2} = \frac{(n+1)n(2n+1)}{6}. \tag{2.31}$$

It is of interest to note that

$$\sum_{k=0}^{n} k = 1 + 2 + \cdots + n$$
$$= n + (n-1) + \cdots + 1.$$

Adding the right-hand sides of the two expressions above and dividing by two yields

$$\sum_{k=0}^{n} k = \frac{n(n+1)}{2}.$$

Gauss discovered this result at a very tender age.

Example 2.3.3. *The discrete RV x has PMF*

$$p_x(\alpha) = \begin{cases} a\alpha, & \alpha = 1, 2, \ldots, 10 \\ 0, & \text{otherwise.} \end{cases}$$

Find: (a) the constant a, (b) the CDF F_x, (c) $P(1 < x)$.

Solution

(a) We have

$$1 = \sum_{\alpha} p_x(\alpha) = a \sum_{\alpha=1}^{10} \alpha = a \frac{11 \cdot 10}{2} = 55a,$$

so that $a = 1/55$.

(b) We find

$$F_x(\alpha) = \sum_{\alpha' \leq \alpha} p_x(\alpha') = \begin{cases} 0, & \alpha < 1 \\ \dfrac{a(k+1)k}{2}, & k \leq \alpha < k+1, K = 1, 2, \ldots, 9 \\ 1, & 10 \leq \alpha. \end{cases}$$

(c) We have

$$P(1 < x) = 1 - P(x \leq 1) = 1 - F_x(1) = 1 - a = \frac{54}{55}.$$

■

Another frequently useful result is the expression for the sum of a geometric series.

Lemma 2.3.2 (Sum of Geometric Series). *Define*

$$S_{m,n}(w) = \sum_{k=m}^{n} w^k,$$

where w is any complex number. Then

$$S_{m,n}(w) = \begin{cases} \dfrac{w^m - w^{n+1}}{1 - w}, & \text{if } n \geq m \text{ and } w \neq 1 \\ n - m + 1, & \text{if } n \geq m \text{ and } w = 1 \\ 0, & \text{if } n < m. \end{cases}$$

Proof. Assume $n \geq m$. We have

$$S_{m,n}(w) = w^m + w^{m+1} + \cdots + w^n,$$

and

$$w\,S_{m,n}(w) = w^{m+1} + w^{m+2} + \cdots + w^n + w^{n+1},$$

so that

$$(1 - w)\,S_{m,n}(w) = w^m - w^{n+1},$$

from which the desired result follows. ∎

Example 2.3.4. *The discrete RV x has PMF*

$$p_x(\alpha) = \begin{cases} a(0.9)^\alpha, & \alpha = 2, 3, 4, \ldots \\ 0, & \text{otherwise.} \end{cases}$$

Find the CDF $F_x(\alpha)$ in closed form, and find the constant a.

Solution. Using the sum of a geometric series,

$$F_x(k) = a\sum_{i=2}^{k}(0.9)^i = \alpha\frac{(0.9)^2 - (0.9^{k+1})}{1 - 0.9}, \quad k = 2, 3, 4, \ldots,$$

so that

$$F_x(\alpha) = \begin{cases} 0, & \alpha < 2 \\ 8.1a(1 - (0.9)^{k-1}), & k \leq \alpha < k + 1, k = 2, 3, \ldots. \end{cases}$$

Since $F_x(\infty) = 1$, we find $a = 10/81$. ∎

2.3.2 Continuous Random Variables

Continuous random variables take on a continuum of values. The resulting CDF is a continuous function. Probabilities for continuous random variables are often easily found with the aid of the probability density function—which can be found from the CDF.

Definition 2.3.5. *A RV x defined on $(S, \mathcal{F}, P)$ is **continuous** if the CDF F_x is absolutely continuous. To avoid technicalities, we simply note that if F_x is absolutely continuous then F_x is continuous everywhere and F_x is differentiable except perhaps at isolated points. Consequently, there exists a function f_x satisfying*

$$F_x(\alpha) = \int_{-\infty}^{\alpha} f_x(\alpha')\, d\alpha'. \tag{2.32}$$

*The function f_x is called the **probability density function** (PDF) for the continuous RV x. The set of points for which the PDF is nonzero is called the **support set** for f_x.*

Theorem 2.3.3. *Let x be a continuous RV. The PDF f_x satisfies*

$$f_x(\alpha) = \frac{d F_x(\alpha)}{d\alpha} \geq 0 \tag{2.33}$$

(except perhaps at isolated points),

$$\int_{-\infty}^{\infty} f_x(\alpha)\, d\alpha = 1, \tag{2.34}$$

and (for any Borel set A)

$$P(\{\zeta : x(\zeta) \in A\}) = \int_A f_x(\alpha')\, d\alpha. \tag{2.35}$$

Proof. We have

$$\frac{d F_x(\alpha)}{d\alpha} = \lim_{h \leftarrow 0} \frac{F_x(\alpha) - F_x(\alpha - h)}{h}$$

$$= \lim_{h \leftarrow 0} \frac{1}{h} \int_{\alpha-h}^{\alpha} f_x(\alpha')\, d\alpha'$$

$$= f_x(\alpha) \lim_{h \leftarrow 0} \frac{1}{h} \int_{\alpha-h}^{\alpha} d\alpha'$$

$$= f_x(\alpha).$$

The above is a special case of Leibnitz' rule.

The PDF f_x is nonnegative since the CDF F_x is monotone nondecreasing.

Since $F_x(+\infty) = 1$, we have

$$1 = \int_{-\infty}^{\infty} f_x(\alpha')\, d\alpha'.$$

From the properties of a CDF,

$$P(a < x \le b) = \int_a^b f_x(\alpha')\, d\alpha';$$

This is easily extended to any Borel set to yield (2.35).   ∎

We consider $f_x(\alpha)$ to be the left-hand derivative of the CDF:

$$f_x(\alpha) = \lim_{h \to 0} \frac{F_x(\alpha) - F_x(\alpha - h)}{h}, \qquad (2.36)$$

where the limit is through positive values of h. For a continuous RV, the right- and left-hand derivatives are equal almost everywhere; however, a treatment of discrete and mixed RVs using a PDF containing Dirac delta functions can be developed. Such a treatment is offered in the following section, with the Dirac delta function

$$\delta(\alpha) = \lim_{h \to 0} \frac{u(\alpha) - u(\alpha - h)}{h}, \qquad (2.37)$$

where the limit requires a special interpretation.

Example 2.3.5.   *The PDF for the RV x is*

$$f_x(\alpha) = \begin{cases} \beta\alpha^2, & -1 < \alpha < 2 \\ 0, & \text{otherwise.} \end{cases}$$

Find β so that f_x is a PDF, and find the CDF F_x.

Solution.   We require

$$1 = \int_{-\infty}^{\infty} f_x(\alpha)\, d\alpha = \beta \int_{-1}^{2} \alpha^2\, d\alpha = \frac{\beta}{3}(8 + 1) = 3\beta$$

so that $\beta = 1/3$. We note that $f_x(\alpha) \ge 0$, as required. We obtain the CDF using

$$F_x(\alpha) = \int_{-\infty}^{\alpha} f_x(\alpha')\, d\alpha'.$$

Since $f_x(\alpha') = 0$ for $\alpha' < -1$ we obtain $F_x(\alpha) = 0$ for $\alpha < -1$. For $-1 \le \alpha < 2$ we obtain

$$F_x(\alpha) = \int_{-1}^{\alpha} \frac{1}{3}\alpha'^2\, d\alpha' = \frac{1}{9}(\alpha^3 + 1).$$

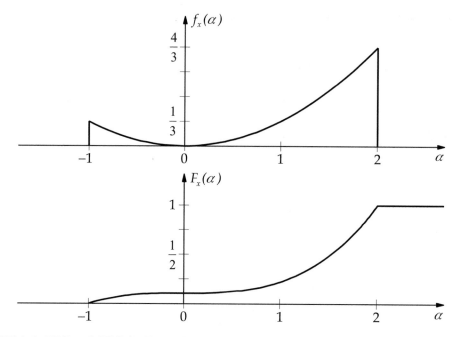

FIGURE 2.5: PDF and CDF for Example 2.3.5

Finally, since $f_x(\alpha') = 0$ for $\alpha' > 2$ we have $F_x(\alpha) = 1$ for $\alpha \geq 2$. Note that $f_x(-1)$ and $f_x(2)$ could be defined to be any real numbers without affecting the result. In fact, the PDF could be redefined at any discrete set of points without affecting the result.

The PDF and CDF for this example are illustrated in Fig. 2.5. It is extremely important to be able to visualize the relationship between the PDF and CDF. ∎

Example 2.3.6. *The RV x has PDF $f_x(\alpha) = 5e^{-5\alpha}u(\alpha)$. Find $P(-1 < x < 2)$.*

Solution. We have

$$P(-1 < x < 2) = \int_{-1}^{2} f_x(\alpha)\, d\alpha = \int_{0}^{2} 5e^{-5\alpha}\, d\alpha = -e^{-5\alpha}\Big|_{0}^{2} = 1 - e^{-10}.$$

2.3.3   Mixed Random Variables

Mixed random variables are neither discrete nor continuous. The resulting CDF is piecewise continuous. Probabilities for mixed random variables can be found in several ways. Splitting the CDF into the sum of a jump CDF and a continuous CDF and using the corresponding PMF and PDF is one approach, and is treated in this section.

Definition 2.3.6. *A RV x defined on $(S, \mathcal{F}, P)$ is a **mixed** RV if it is neither discrete nor continuous.*

Theorem 2.3.4 (Lebesgue Decomposition Theorem). *A CDF F may be expressed as*

$$F(\alpha) = \gamma F_C(\alpha) + (1 - \gamma) F_D(\alpha), \qquad (2.38)$$

where $0 \le \gamma \le 1$, F_C is a continuous CDF, and F_D is a discrete CDF.

Proof. Note that if F is continuous then $\gamma = 1$ and $F_C = F$. Similarly, if F is discrete then $\gamma = 0$ and $F_D = F$.

Assume F is neither continuous nor discrete. Define

$$q(\alpha) = F(\alpha) - F(\alpha^-).$$

Then $q(\alpha) \ge 0$, and $q(\alpha) \ne 0$ only at isolated points, say $\alpha \in D$. Let

$$(1 - \gamma) F_D(\alpha) = \sum_{\alpha' \le \alpha} q(\alpha')$$

and

$$1 - \gamma = \sum_{\alpha' \in D} q(\alpha').$$

Then F_D is a monotone nondecreasing, right-continuous jump function, with $F_D(-\infty) = 0$ and $F_D(+\infty) = 1$; i.e., F_D is a discrete CDF. Now, let

$$F_C(\alpha) = \frac{F(\alpha) - (1 - \gamma) F_D(\alpha)}{\gamma}.$$

Then $F_C(-\infty) = 0$, $F_C(+\infty) = 1$, and F_C is right-continuous since both F and F_D are right-continuous. Also,

$$F_C(\alpha) - F_C(\alpha^-) = \frac{q(\alpha) - q(\alpha)}{\gamma} = 0.$$

Consequently, F_C is a continuous CDF. ∎

Although the above decomposition theorem is useful, there is no guarantee that F_C is absolutely continuous and hence that F_C is the CDF of a continuous RV. It can be shown that F_C can always be further decomposed into the sum of an absolutely continuous part and a singular part [4]. We will assume throughout that the singular part is zero, and hence that F_C is the CDF for a continuous RV. All CDFs arising in practical applications satisfy this assumption. For our purposes then, if $\gamma = 1$ the CDF describes a continuous RV and if $\gamma = 0$ the CDF describes a discrete RV.

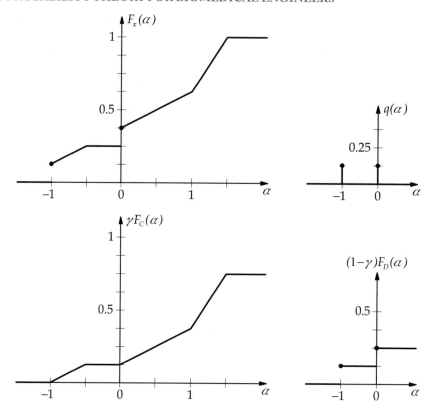

FIGURE 2.6: Plots for Example 2.3.7

Example 2.3.7.   *The RV x has CDF*

$$
F_x(\alpha) = \begin{cases}
0, & \alpha < -1 \\[6pt]
\dfrac{1}{8} + \dfrac{1}{4}(\alpha + 1), & -1 \le \alpha < -\dfrac{1}{2} \\[6pt]
\dfrac{1}{4}, & -\dfrac{1}{2} \le \alpha < 0 \\[6pt]
\dfrac{3}{8} + \dfrac{1}{4}\alpha, & 0 \le \alpha < 1 \\[6pt]
\dfrac{5}{8} + \dfrac{3}{4}(\alpha - 1), & 1 \le \alpha < \dfrac{3}{2} \\[6pt]
1, & \dfrac{3}{2} \le \alpha.
\end{cases}
$$

Express F_x as $F_x = \gamma F_C + (1 - \gamma)F_D$, where F_C is a continuous CDF and F_D is a discrete CDF.

Solution.   Plots for this example are given in Fig. 2.6. Following the notation in the proof of

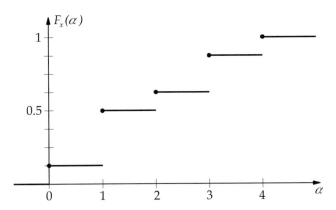

FIGURE 2.7: Cumulative distribution function for Drill Problem 2.3.1

the Lebesgue Decomposition Theorem,

$$q(\alpha) = F_x(\alpha) - F_x(\alpha^-) = \begin{cases} \dfrac{1}{8}, & \alpha = -1, 0 \\ 0, & \text{otherwise.} \end{cases}$$

Hence,

$$(1-\gamma)F_D(\alpha) = \sum_{\alpha' \le \alpha} q(\alpha') \begin{cases} 0 & \alpha < -1 \\ \dfrac{1}{8}, & -1 \le \alpha < 0 \\ \dfrac{1}{4}, & 0 \le \alpha; \end{cases}$$

so that $1 - \gamma = 1/4$ and $\gamma = 3/4$. Finally, $\gamma F_C = F_x - (1-\gamma)F_D$, or

$$\gamma F_C(\alpha) = \begin{cases} 0, & \alpha < -1 \\ \dfrac{1}{4}(\alpha + 1), & -1 \le \alpha < -\dfrac{1}{2} \\ \dfrac{1}{8}, & -\dfrac{1}{2} \le \alpha < 0 \\ \dfrac{1}{8} + \dfrac{1}{4}\alpha, & 0 \le \alpha < 1 \\ \dfrac{3}{8} + \dfrac{3}{4}(\alpha - 1), & 1 \le \alpha < \dfrac{3}{2} \\ \dfrac{3}{4}, & \dfrac{3}{2} \le \alpha. \end{cases}$$

∎

Drill Problem 2.3.1. *The discrete random variable x has cumulative distribution function F_x shown in Fig. 2.7. Find (a) $p_x(-1)$, (b) $p_x(0)$, (c) $P(0 \le x \le 3)$, (d) $P(0 < x \le 2)$.*

Answers: 7/8, 1/8, 0, 1/2.

Drill Problem 2.3.2. *A committee of three members is to be formed from four engineers and three physicists. Let x be a RV which assigns to every sample point in S a value equal to the number of engineers on the committee. Determine: (a) $p_x(0)$, (b) $p_x(1)$, (c) $p_x(2)$, and (d) $p_x(3)$.*

Answers: 1/35, 4/35, 12/35, 18/35.

Drill Problem 2.3.3. *The waveform $w(t)$ is uniformly sampled every 0.2 s from $t = 0$ s to $t = 4$ s, where*

$$w(t) = \begin{cases} 2t, & 0 \leq t \leq 2 \\ 4e^{-2(t-2)}, & 2 < t \leq 4. \end{cases}$$

The sampled values are rounded off to the nearest integer and collected in the set S. The RV $x(\zeta) = \zeta$ for all $\zeta \in S$. Determine: (a) $p_x(0)$, (b) $p_x(1)$, (c) $p_x(2)$, (d) $p_x(3)$, (e) $p_x(5)$.

Answers: 1/3, 5/21, 4/21, 1/7, 0.

Drill Problem 2.3.4. *Suppose the RV x has the PMF*

$$p_x(\alpha) = \begin{cases} \dfrac{1}{8}, & \alpha = 0, 2 \\ \dfrac{1}{4}, & \alpha = 1, 3, 4 \\ 0, & \text{otherwise.} \end{cases}$$

Find: (a) $F_x(-1)$, (b) $F_x(0)$, (c) $F_x(1)$, (d) $F_x(3)$.

Answers: 3/8, 1/8, 3/4, 0.

Drill Problem 2.3.5. *The PDF for the RV x is*

$$f_x(\alpha) = \begin{cases} \beta(\alpha + 1), & -1 < \alpha < 2 \\ 0, & \text{otherwise,} \end{cases}$$

where β is a constant. Determine: (a) β, (b) $P(x \leq -1)$, (c) $F_x(0)$, (d) $P(0 \leq x \leq 2)$.

Answers: 0, 1/9, 2/9, 8/9.

Drill Problem 2.3.6. *The PDF for the RV x is*

$$f_x(\alpha) = \begin{cases} \beta(\alpha^{1/2} + \alpha^{-1/2}), & 0 < \alpha < 1 \\ 0, & \text{otherwise,} \end{cases}$$

where β is a constant. Determine: (a) β, (b) $P(x \geq 1/2)$, (c) $F_x(1/4)$, (d) $P(x = 1/4)$.

Answers: 0.406, 3/8, 0, 0.381.

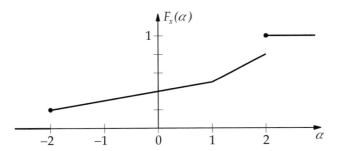

FIGURE 2.8: Cumulative distribution function for Drill Problem 2.3.8

Drill Problem 2.3.7. *The CDF for the RV x is*

$$F_x(\alpha) = \begin{cases} 0, & \alpha < -1 \\ 3(\alpha - \alpha^3/3 + 2/3)/4, & -1 \le \alpha < 1 \\ 1, & 1 \le \alpha. \end{cases}$$

Determine: (a) $F_x(0)$, *(b)* $P(x \ge 1/2)$, *(c)* $f_x(0)$, *(d)* $f_x(4/3)$.

Answers: 5/32, 1/2, 0, 3/4.

Drill Problem 2.3.8. *Random variable x has the mixed CDF F_x shown in Fig. 2.8. Find (a)* $P(-1 \le x < 0.5)$, *(b)* $P(-2 < x < -1)$, *(c)* $P(-2 \le x < -1)$, *(d)* $P(x > 1.5)$.

Answers: 0.1, 0.15, 0.35, 0.3.

2.4   RIEMANN-STIELTJES INTEGRATION

We will have a great interest in evaluating integrals of the form

$$\int_a^b dF(\alpha),$$

$$\int_B dF(\alpha),$$

and

$$\int_a^b g(\alpha) \, dF(\alpha),$$

where F is a CDF, a and b are real numbers, B is a Borel set, and $g : R^* \mapsto R^*$. Such integrals are known as Riemann-Stieltjes integrals. In the following we assume that F is the CDF for the RV x and that $a < b$. In the special case that $B = (a, b]$ and $g(\alpha) = 1$ for all α, then all the above integrals are the same. We establish below that

$$P(\{\zeta : a < x(\zeta) \le b\}) = F(b) - F(a) = \int_a^b d F(\alpha). \qquad (2.39)$$

The Riemann-Stieltjes integral provides a unified framework for treating continuous, discrete, and mixed RVs—all with one kind of integration. An important alternative is to use a standard Riemann integral for continuous RVs, a summation for discrete RVs, and a Riemann integral with an integrand containing Dirac delta functions for mixed RVs.

We begin with a brief review of the standard Riemann integral. Let

$$a = \alpha_0 < \alpha_1 < \alpha_2 < \cdots < \alpha_n = b, \qquad (2.40)$$

$$\alpha_{i-1} \le \xi_i \le \alpha_i, \quad i = 1, 2, \ldots, n, \qquad (2.41)$$

and

$$\Delta_n = \max_{1 \le i \le n} \{\alpha_i - \alpha_{i-1}\}. \qquad (2.42)$$

The **Riemann integral** is defined by

$$\int_a^b h(\alpha) \, d\alpha = \lim_{\Delta_n \to 0} \sum_{i=1}^n h(\xi_i)(\alpha_i - \alpha_{i-1}), \qquad (2.43)$$

provided the limit exists and is independent of the choice of $\{\xi_i\}$. Note that $n \to \infty$ as $\Delta_n \to 0$. The summation above is called a Riemann sum. We remind the reader that this is the "usual" integral of calculus and has the interpretation as the area under the curve h between a and b.

With the same notation as above, the **Riemann-Stieltjes integral** is defined by

$$\int_a^b g(\alpha) \, d F(\alpha) = \lim_{\Delta_n \to 0} \sum_{i=1}^n g(\xi_i)(F(\alpha_i) - F(\alpha_{i-1})), \qquad (2.44)$$

provided the limit exists and is independent of the choice of $\{\xi_i\}$.

Applying the above definition, we obtain (as promised)

$$\int_a^b dF(\alpha) = \lim_{\Delta_n \to 0} ((F(\alpha_1) - F(\alpha_0)) + (F(\alpha_2) - F(\alpha_1)) + \cdots + (F(\alpha_n) - F(\alpha_{n-1})))$$
$$= F(b) - F(a).$$

Suppose F is discrete with jumps at $\beta \in \{\beta_0, \beta_1, \ldots, \beta_N\}$ satisfying

$$a = \beta_0 < \beta_1 < \beta_2 < \cdots < \beta_N \le b. \tag{2.45}$$

Then, provided that g and F have no common points of discontinuity, it is easily shown that

$$\int_a^b g(\alpha)\, dF(\alpha) = \sum_{i=1}^N g(\beta_i)(F(\beta_i) - F(\beta_i^-)) = \sum_{i-1}^N g(\beta_i) p(\beta_i), \tag{2.46}$$

where $p(\beta) = F(\beta) - F(\beta^-)$. Note that a jump in F at a is not included in the sum whereas a jump at b is included.

Suppose F is absolutely continuous with

$$f(\alpha) = \frac{dF(\alpha)}{d\alpha}. \tag{2.47}$$

Then

$$\int_a^b g(\alpha)\, dF(\alpha) = \int_a^b g(\alpha) f(\alpha)\, d\alpha. \tag{2.48}$$

Hence, the Riemann-Stieltjes integral reduces to the usual Riemann integral in this case.

Defining

$$\int_B dF(\alpha) = P(x^{-1}(B)), \tag{2.49}$$

we find that if $B = (a, b]$ then

$$\int_B dF(\alpha) = \int_a^b dF(\alpha). \tag{2.50}$$

The above summary of Riemann-Stieltjes integration together with the Lebesgue Decomposition Theorem provides a powerful technique for evaluating the integrals encountered in probability theory. With

$$F(\alpha) = \gamma F_C(\alpha) + (1 - \gamma)F_D(\alpha), \tag{2.51}$$

$$f_C(\alpha) = \frac{d F_C(\alpha)}{d\alpha}, \tag{2.52}$$

$$p(\alpha) = F_D(\alpha) - F_D(\alpha^-), \tag{2.53}$$

and

$$D_x = \{\alpha : p(\alpha) \neq 0\}, \tag{2.54}$$

we obtain

$$\int_a^b g(\alpha)d F(\alpha) = \int_a^b g(\alpha)\gamma f_c(\alpha)d\alpha + \sum_{\alpha \epsilon (a,b) \cap D} g(\alpha)(1 - \gamma)p(\alpha). \tag{2.55}$$

The evaluation of the above Riemann-Stieltjes integral is even further simplified by noting that

$$(1 - \gamma)p(\alpha) = F(\alpha) - F(\alpha^-) \tag{2.56}$$

and that

$$\gamma f_C(\alpha) = \frac{d F(\alpha)}{d\alpha}, \quad \text{wherever } p(\alpha) = 0. \tag{2.57}$$

Example 2.4.1. *The RV x has CDF*

$$F_x(\alpha) = \begin{cases} 0, & \alpha < -3 \\[4pt] \dfrac{1}{4}, & -3 \le \alpha < -2 \\[8pt] \dfrac{1}{4} + \dfrac{1}{4}(\alpha + 2), & -2 \le \alpha < -1 \\[8pt] \dfrac{1}{2}, & -1 \le \alpha < 0 \\[8pt] \dfrac{5}{8} + \dfrac{3}{8}\alpha^2, & 0 \le \alpha < 1 \\[8pt] 1, & 1 \le \alpha. \end{cases}$$

Evaluate

$$\int_{-\infty}^{\infty} \alpha^2 \, d F_x(\alpha).$$

Solution. We have $D_x = \{-3, 0\}$,

$$(1 - \gamma)p(\alpha) = \begin{cases} \dfrac{1}{4}, & \alpha = -3 \\[8pt] \dfrac{1}{8}, & \alpha = 0 \\[6pt] 0, & \text{otherwise,} \end{cases}$$

and

$$\gamma f_C(\alpha) = \begin{cases} \dfrac{1}{4}, & -2 < \alpha < -1 \\ \dfrac{3}{4}\alpha, & 0 < \alpha < 1 \\ 0, & \text{otherwise.} \end{cases}$$

Consequently,

$$\int_{-\infty}^{\infty} \alpha^2 \, dF_x(\alpha) = \frac{1}{4}\int_{-2}^{-1}\alpha^2 \, d\alpha + \frac{3}{4}\int_{0}^{1}\alpha^3 \, d\alpha + (-3)^2\frac{1}{4} = \frac{145}{48}.$$ ∎

Example 2.4.2. *Let $F(\alpha) = 0.5\alpha(u(\alpha) - u(\alpha - 1)) + u(\alpha - 1)$. Evaluate*

$$I = \int_{-\infty}^{\infty} \alpha \, dF(\alpha).$$

Solution. We find

$$I = \int_{0}^{1} \frac{1}{2}\alpha \, d\alpha + 1 \cdot \frac{1}{2} = \frac{1}{4} + \frac{1}{2} = \frac{3}{4}.$$ ∎

The Dirac delta function provides an alternative technique for evaluating the integrals occurring in the applications of probability theory.

Definition 2.4.7. *We say that $\delta(\cdot)$ is a **Dirac delta function** if*

$$\int_{-\infty}^{\infty} g(\alpha)\delta(\alpha - \alpha_0) \, d\alpha = g(\alpha_0) \qquad (2.58)$$

for each function $g(\alpha)$ which is continuous at $\alpha = \alpha_0$.

For example, let

$$g(\alpha) = \begin{cases} 1, & |\alpha - \alpha_0| < \varepsilon \\ 0, & \text{otherwise.} \end{cases}$$

Then for all $\varepsilon > 0$, $g(\alpha)$ is continuous at $\alpha = \alpha_0$ and

$$g(\alpha_0) = 1 = \int_{-\infty}^{\infty} g(\alpha)\delta(\alpha - \alpha_0) \, d\alpha = \int_{-\varepsilon}^{\varepsilon} \delta(\alpha') \, d\alpha'.$$

Consequently, $\delta(\alpha)$ has unit area and (virtually) zero width. We conclude that $\delta(0) = \infty$.

Formally, we may treat the Dirac delta function as the derivative of the unit step function,

$$\delta(\alpha) = \frac{du(\alpha)}{d\alpha}, \qquad (2.59)$$

since

$$\int_{-\infty}^{\infty} g(\alpha)\, du(\alpha - \alpha_0) = g(\alpha_0). \qquad (2.60)$$

Letting $\alpha' - \alpha_0 = \alpha_0 - \alpha$, we have $d\alpha' = -d\alpha$ and

$$\int_{-\infty}^{\infty} g(\alpha)\delta(\alpha_0 - \alpha)\, d\alpha = -\int_{\infty}^{-\infty} g(2\alpha_0 - \alpha')\delta(\alpha' - \alpha_0)\, d\alpha' = g(\alpha_0).$$

Consequently, we may treat the Dirac delta function as an even function:

$$\delta(-\alpha) = \delta(\alpha). \qquad (2.61)$$

Similarly, we can easily show that if $g(\alpha)$ is continuous at $\alpha = \alpha_0$, then we have

$$g(\alpha)\delta(\alpha - \alpha_0) = g(\alpha_0)\delta(\alpha - \alpha_0). \qquad (2.62)$$

Example 2.4.3. *Evaluate the following integrals:*

$$I_1 = \int_{-\infty}^{\infty} e^{-\alpha/2}\delta(\alpha - 2)\, d\alpha,$$

$$I_2 = \int_{-\infty}^{0} e^{-\alpha/2}\delta(\alpha - 2)\, d\alpha,$$

$$I_3 = \int_{-\infty}^{\infty} e^{-|\alpha|}\delta(2\alpha + 4)\, d\alpha,$$

$$I_4 = \int_{-\infty}^{\infty} \frac{5\tan(2\alpha) + 3\alpha^2}{\cos(5\alpha - 2) + \sin(\alpha)}\delta(\alpha + 2)\, d\alpha,$$

$$I_5 = \int_{-\infty}^{\infty} (\alpha - 5)(3\delta(\alpha + 3) - 2\delta(\alpha - 2))\, d\alpha,$$

and

$$I_6 = \int\limits_0^3 (\alpha - 5)(3\delta(\alpha + 3) - 2\delta(\alpha - 2)) \, d\alpha \, .$$

Solution. We have $I_1 = e^{-2/2} = e^{-1}$. $I_2 = 0$ since the integration interval does not include $\alpha = 2$. Letting $\alpha' = 2\alpha$ in I_3,

$$I_3 = \frac{1}{2} \int\limits_{-\infty}^\infty e^{-|\alpha'/2|} \delta(\alpha' + 4) \, d\alpha' = \frac{1}{2} e^{-2}.$$

Evaluating I_4,

$$I_4 = \frac{5 \tan(-4) + 3 \cdot 4}{\cos(-12) + \sin(-2)} = -94.90.$$

Now $I_5 = 3(-3 - 5) - 2(2 - 5) = -18$ and $I_6 = -2(2 - 5) = 6$. ∎

By allowing Dirac delta functions, we may let

$$f(\alpha) = \frac{dF(\alpha)}{d\alpha} \tag{2.63}$$

to obtain

$$\int\limits_a^b g(\alpha) \, dF(\alpha) = \int\limits_a^b g(\alpha) f(\alpha) \, d\alpha \, . \tag{2.64}$$

Extreme caution must be used in interpreting the latter integral when F contains a jump at either a (which should not be included) or at b (which should be included). In particular, since

$$F(\alpha) = \int\limits_{-\infty}^\alpha dF(\alpha') \tag{2.65}$$

and F is right-continuous, we must use care when evaluating

$$F(\alpha) = \int_{-\infty}^\alpha f(\alpha') d\alpha' \tag{2.66}$$

if f contains Dirac delta functions.

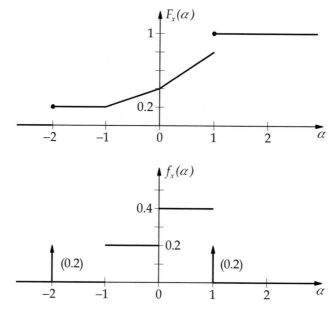

FIGURE 2.9: Cumulative distribution function and probability density function for Example 2.4.4

Example 2.4.4. *Random variable x has CDF F_x given by*

$$F_x(\alpha) = \begin{cases} 0, & \alpha < -2 \\ 0.2, & -2 \leq \alpha < -1 \\ 0.2 + 0.2(\alpha + 1), & -1 \leq \alpha < 0 \\ 0.4 + 0.4\alpha, & 0 \leq \alpha < 1 \\ 1, & \alpha \geq 1. \end{cases}$$

Sketch F_x. Find and sketch the PDF f_x.

Solution. Note that each piece of the given CDF is continuous. Examining the endpoints of each interval reveals that the CDF has discontinuities at $\alpha = -2$ and at $\alpha = 1$. The CDF and PDF are shown in Fig. 2.9. As is common practice, we have shown the Dirac delta functions as arrows with length corresponding to the area under the delta function. In addition, the weight (area) is shown next to each delta function. The PDF may be expressed as

$$f_x(\alpha) = 0.2\delta(\alpha + 2) + 0.2u(\alpha + 1) + 0.2u(\alpha) - 0.4u(\alpha - 1) + 0.2\delta(\alpha - 1).$$

The CDF may be expressed as

$$F_x(\alpha) = 0.2u(\alpha + 2) + 0.2(\alpha + 1)(u(\alpha + 1) - u(\alpha))$$
$$+ (0.2 + 0.4\alpha)(u(\alpha) - u(\alpha - 1)) + 0.8u(\alpha - 1).$$

The reader is encouraged to differentiate the above expression for F_x to obtain f_x. It should be apparent that plotting the CDF and PDF significantly reduces the work involved.   ■

Example 2.4.5. *A coin is tossed n times. The probability of a head on any one toss is p and the probability of a tail is q, where p + q = 1. Let the RV x be the number of heads in n tosses. Find the CDF and the PDF for the RV x.*

Solution. The PMF for x is

$$p_x(\alpha) = \begin{cases} \binom{n}{k} p^k q^{n-k}, & \alpha = k = 0, 1, \ldots, n \\ 0, & \text{otherwise.} \end{cases}$$

Consequently, the CDF is

$$F_x(\alpha) = \sum_{k=0}^{N} \binom{n}{k} p^k q^{n-k} u(\alpha - k)$$

and the PDF is

$$F_x(\alpha) = \sum_{k=0}^{n} \binom{n}{k} p^k q^{n-k} \delta(\alpha - k).$$ ■

Drill Problem 2.4.1. *The CDF of the RV x is given by*

$$F_x(\alpha) = \begin{cases} 0, & \alpha < -2 \\ \dfrac{1}{4} + \dfrac{1}{12}(\alpha + 2), & -2 \le \alpha < 1 \\ \dfrac{1}{2} + \dfrac{1}{4}(\alpha - 1), & 1 \le \alpha < 2 \\ 1, & \alpha \ge 2. \end{cases}$$

Evaluate

$$I_1 = \int_{-\infty}^{\infty} \alpha \, d F_x(\alpha)$$

and

$$I_2 = \int_{-\infty}^{\infty} \alpha^2 \, d F_x(\alpha).$$

Answers: 1/4, 17/6.

Drill Problem 2.4.2.   *Evaluate the following integrals:*

$$I_1 = \int_{-\infty}^{\infty} \ln(\sin(\pi\alpha)) \, du(\alpha - 0.3);$$

$$I_2 = \int_{-\infty}^{\infty} \sin(\pi\alpha)(du(\alpha) + du(\alpha - 2));$$

$$I_3 = \int_{-\infty}^{\infty} 2\alpha \, du(\alpha + 3);$$

and

$$I_4 = \int_{-\infty}^{\infty} 5u(t - 3) \, du(t + 3).$$

Answers: 0, −6, 0, −0.212.

Drill Problem 2.4.3.   *Evaluate the following integrals:*

$$I_1 = \int_{-\infty}^{\infty} \ln(\sin(\pi\alpha))\delta(\alpha - 0.3) \, d\alpha;$$

$$I_2 = \int_{-\infty}^{\infty} \sin(\pi\alpha)(\delta(\alpha) + \delta(\alpha - 2)) \, d\alpha;$$

$$I_3 = \int_{-\infty}^{\infty} 2\alpha\delta(\alpha + 3) \, d\alpha;$$

and

$$I_4 = \int_{-\infty}^{\infty} 5u(t - 3)\delta(t + 3) \, dt.$$

Answers: 0, −6, 0, −0.212.

Drill Problem 2.4.4.   *Two balls are selected at random from an urn that contains two blue, three red, and three green balls. Find the PDF for the random variable x, where x is the number of blue balls selected.*

Answer: $f_x(\alpha) = \dfrac{1}{28}\delta(\alpha - 2) + \dfrac{3}{7}\delta(\alpha - 1) + \dfrac{15}{28}\delta(\alpha).$

2.5   CONDITIONAL PROBABILITY

In Chapter 1, we discussed conditional probabilities. With events A and B defined on the probability space $(S, \mathcal{F}, P)$, we defined the probability that event B occurs, given that event A occurred as

$$P(B|A) = \frac{P(A \cap B)}{P(A)}. \qquad (2.67)$$

Definition 2.5.1. *Let x be a RV defined on $(S, \mathcal{F}, P)$, and let B denote the event*

$$B = \{\zeta : x(\zeta) \leq \alpha\}.$$

*The **conditional CDF** for the RV x, given event A, is defined by*

$$F_{x|A}(\alpha|A) = \frac{P(A \cap B)}{P(A)} = \frac{P(\{\zeta \in S : x(\zeta) \leq \alpha, \zeta \in A\})}{P(A)}. \qquad (2.68)$$

If $P(A) = 0$ we define $F_{x|A}$ to be any valid CDF.

*   If x is a discrete RV, we define the **conditional PMF** for the RV x, given event A, by*

$$p_{x|A}(\alpha|A) = F_{x|A}(\alpha|A) - F_{x|A}(\alpha^-|A). \qquad (2.69)$$

*Similarly, if x is a continuous RV, we define the **conditional PDF** x, given event A, by*

$$f_{x|A}(\alpha|A) = \frac{d F_{x|A}(\alpha|A)}{d\alpha}. \qquad (2.70)$$

Note that the conditional CDF $F_{x|A}$ is indeed a CDF in its own right; i.e., $F_{x|A}$ is monotone nondecreasing, right-continuous, $F_{x|A}(-\infty|A) = 0$, and $F_{x|A}(\infty|A) = 1$.

    If x is a discrete RV and $P(A) \neq 0$, from (2.69) we have

$$p_{x|A}(\alpha|A) = \frac{P(\zeta \in S : x(\zeta) = \alpha, \zeta \in A)}{P(A)},$$

so that

$$p_{x|A}(\alpha|A) = \begin{cases} \dfrac{p_{x|A}(\alpha)}{P(A)}, & x^{-1}(\{\alpha\}) \subset A \\ 0, & \text{otherwise.} \end{cases} \qquad (2.71)$$

Similarly, if x is a continuous RV and $P(A) \neq 0$, it follows from (2.70) that

$$f_{x|A}(\alpha|A) = \begin{cases} \dfrac{f_x(\alpha)}{P(A)}, & x^{-1}(\{\alpha\}) \subset A \\ 0, & \text{otherwise.} \end{cases} \qquad (2.72)$$

Recall the discussion in Section 1.8 that a probability space $(A, \mathcal{F}_A, P_A)$ can be defined such that all conditional probabilities (given event A) on $(S, \mathcal{F}, P)$ may be computed as unconditional probabilities on $(A, \mathcal{F}_A, P_A)$. Consequently, all remarks and properties regarding a CDF, PMF, and PDF are also valid for the corresponding conditional entities. On the probability space $(A, \mathcal{F}_A, P_A)$ we may define the RV $y = x|A$ with CDF $F_y(\alpha) = F_{x|A}(\alpha|A)$.

Example 2.5.1. *Let the RV x have the PDF*

$$f_x(\alpha) = \begin{cases} 1 - \alpha, & 0 < \alpha < 1 \\ \alpha - 1, & 1 < \alpha < 2 \\ 0, & \text{otherwise.} \end{cases}$$

Define the events $A = \{x > 1\}$ and $B = \{0.5 < x < 1.5\}$. Find (a) $F_{x|A}(\alpha|A)$; (b) $f_{x|A}(\alpha|A)$; and (c) $f_{x|B}(\alpha|B)$.

Solution

(a) By definition,

$$F_{x|A}(\alpha|A) = \frac{P(x \leq \alpha, x > 1)}{P(A)} = \frac{P(1 < x \leq \alpha)}{P(x > 1)}.$$

Integrating the PDF from 1 to 2 we find that $P(A) = P(x > 1) = 0.5$, and

$$F_{x|A}(\alpha|A) = \begin{cases} 0, & \alpha < 1 \\ 2(F_x(\alpha) - 0.5) = \alpha^2 - 2\alpha + 1, & 1 \leq \alpha < 2 \\ 1, & \alpha \geq 2. \end{cases}$$

(b) Differentiating the result from (a) we obtain

$$f_{x|A}(\alpha|A) = \begin{cases} 0, & \alpha < 1 \\ 2\alpha - 2, & 1 < \alpha < 2 \\ 0, & \alpha > 2. \end{cases}$$

As an alternative, from (2.72) we obtain

$$f_{x|A}(\alpha|A) = \begin{cases} 2f_x(\alpha) = 2(\alpha - 1), & 1 < \alpha < 2 \\ 0, & \text{otherwise.} \end{cases}$$

(c) From the given PDF we find $P(B) = P(0.5 < x < 1.5) = 0.25$. Applying the definition of conditional CDF, we find

$$F_{x|B}(\alpha|B) = \frac{P(0.5 < x < \min\{\alpha, 1.5\})}{P(B)}.$$

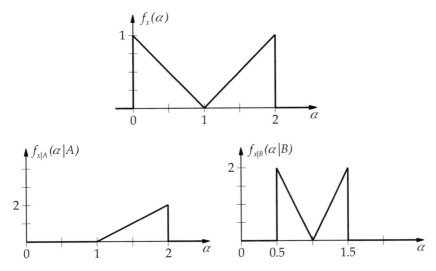

FIGURE 2.10: PDFs for Example 2.5.1

Consequently,

$$f_{x|B}(\alpha|B) = \frac{d F_{x|B}(\alpha|B)}{d\alpha} = \begin{cases} 4(1-\alpha), & 0.5 < \alpha < 1 \\ 4(\alpha-1), & 1 < \alpha < 1.5 \\ 0, & \text{otherwise.} \end{cases}$$

The PDFs f_x, $f_{x|A}$, and $f_{x|B}$ are illustrated in Fig. 2.10. ∎

Example 2.5.2. *The spread of an infection in a family is described by the following PMF*

$$P(S_{n+1} = k | S_n \text{ and } I_n) = \binom{S_n}{k} q_n^k (1 - q_n)^{S_n-k}$$

where n is the sampling interval, S_{n+1} is the number of susceptibles during the next sampling interval and S_n is the number of susceptibles during the current sampling interval, I_n is the number of infectives during the current sampling interval, p is the probability of adequate contact between a susceptible and any infective during one sampling interval, and $q_n = (1 - p)^{I_n}$ is the probability that a susceptible avoids contact with all infectives. This PMF is called the Reed–Frost model and provides the probability of a certain number of susceptibles at a particular sampling interval given a certain number of susceptibles and infectives during the previous sampling internal [2, 10, 14]. If $S_0 = 5$, $I_0 = 1$ and $p = 0.2$, find the probability that three additional family members are infected by the third sampling interval.

Solution. Background for this problem is given in footnote.[1] This problem is most easily visualized using a tree diagram, from which we find

$$
\begin{aligned}
P(S_3 = 2) = {} & P(S_3 = 2|S_1 = 2, \ S_2 = 2) \, P(S_2 = 2|S_1 = 2) \, P(S_1 = 2) \\
& + P(S_3 = 2|S_1 = 3, \ S_2 = 2) \, P(S_2 = 2|S_1 = 3) \, P(S_1 = 3) \\
& + P(S_3 = 2|S_1 = 3, \ S_2 = 3) \, P(S_2 = 3|S_1 = 3) \, P(S_1 = 3) \\
& + P(S_3 = 2|S_1 = 4, \ S_2 = 2) \, P(S_2 = 2|S_1 = 4) \, P(S_1 = 4) \\
& + P(S_3 = 2|S_1 = 4, \ S_2 = 3) \, P(S_2 = 3|S_1 = 4) \, P(S_1 = 4) \\
& + P(S_3 = 2|S_1 = 4, \ S_2 = 4) \, P(S_2 = 4|S_1 = 4) \, P(S_1 = 4) \\
& + P(S_3 = 2|S_1 = 5, \ S_2 = 2) \, P(S_2 = 2|S_1 = 5) \, P(S_1 = 5) \\
& + P(S_3 = 2|S_1 = 5, \ S_2 = 3) \, P(S_2 = 3|S_1 = 5) \, P(S_1 = 5) \\
& + P(S_3 = 2|S_1 = 5, \ S_2 = 4) \, P(S_2 = 4|S_1 = 5) \, P(S_1 = 5) \\
& + P(S_3 = 2|S_1 = 5, \ S_2 = 5) \, P\big(S_2 = 5\big|S_1 = 5\big) \, P(S_1 = 5) \\[4pt]
= {} & 0.64 \times 0.64 \times 0.0512 \\
& + 0.64 \times 0.384 \times 0.2048 \\
& + 0.384 \times 0.512 \times 0.2048 \\
& + 0.64 \times 0.1536 \times 0.4096 \\
& + 0.384 \times 0.4096 \times 0.4096 \\
& + 0.1536 \times 0.4096 \times 0.4096 \\
& + 0.64 \times 0.0512 \times 0.32768 \\
& + 0.384 \times 0.2048 \times 0.32768 \\
& + 0.1536 \times 0.4096 \times 0.32768 \\
& + 0.0512 \times 0.32768 \times 0.32768 \\[4pt]
= {} & 0.006710886 + 0.050331648 + 0.040265318 + 0.040265318 + 0.064424509 \\
& + 0.025769804 + 0.010737418 + 0.025769804 + 0.020615843 + 0.005497558 \\[4pt]
= {} & 0.290388108 \qquad \blacksquare
\end{aligned}
$$

[1] The history of infectious or communicable disease modeling dates to 1760 when D. Bernoulli studied the population dynamics of smallpox with a mathematical model. Little work was done until the early 20th century when Hammer and Soper presented mathematical models which described the spread of measles in Glasgow Scotland. In 1928, Kermack and McKendrick (continuous time), and Reed and Frost (discrete time) presented extensions of the work of Hammer and Soper. Since the 1950s, when Abbey and Bailey presented their work, there has been an epidemic in work in this area.

Infections are spread by adequate contact between two populations, those who are susceptible and those who are infected. The Kermack and McKendrick is a continuous deterministic model that describes the spread of an infection in a large population. The Reed and Frost model is a discrete-time probabilistic model that describes the spread of an infection in a small population. A discrete-time deterministic extension of the Reed and Frost model is useful for exploring the spread of an infection in a large population. One reason for utilizing a discrete-time model rather than a continuous-time model is that *recorded* data is measured at regular intervals. Another reason is that extensions to this model are easily accomplished, such as adding a nonzero latent period with a precisely defined distribution.

Theorem 2.5.1 (Total Probability).   *Let* $\{A_i\}_{i=1}^n$ *be a partition of* S *with* $A_i \in \mathcal{F}$, $i = 1, 2, \ldots, n$, *and let* x *be a RV defined on the probability space* $(S, \mathcal{F}, P)$. *Then*

$$F_x(\alpha) = \sum_{i=1}^n F_{x|A_i}(\alpha|A_i) P(A_i). \tag{2.73}$$

Here we assume:

1. Uniform mixing.
2. Nonzero latent period (the time elapsed between contact and the actual discharge of the infectious agent).
3. Population is closed and at steady state.
4. Any susceptible individual after contact with an infectious person develops the infection, and is infectious to others only in the following period, after which they are immune (immune individuals, R, no longer transmits the agent, and are either temporarily or permanently immune to the disease).
5. Since the person can be infected at any instant during the time period — the average latent period is $1/2$ of the time period, where the length of the time period represents the period of infectivity.
6. Each individual has a fixed probability of coming into adequate contact p with any other specified individual within one time period.

Note that the probability of adequate contact p can be thought of as

$$p = \frac{\text{average number of adequate contacts}}{N}.$$

With

$$q = 1 - p,$$

the probability that a susceptible individual does not come into adequate contact is

$$q^{I_n}.$$

The structure of the Reed-Frost model is shown in the following diagram.

The Reed-Frost model describes the transfer of S susceptibles, I infectives, and R immunes from state to state at sampling interval $n + 1$. After adequate contact with an infective in a given sampling interval, a susceptible will develop the infection, and be infectious to others only during the subsequent sampling interval, and after which, becomes immune.

Since order does not matter when a susceptible individual becomes infected, the number of combinations that S_n survive taken k at a time is $\binom{S_n}{k}$. Then the probability that $S_{n+1} = k$ follows as

$$P(S_{n+1} = k | S_n \text{ and } I_n) \binom{S_n}{k} q_n^k (1 - q_n)^{S_n - k} \text{ for } k = 0, \cdots, S_n.$$

If the RV x is discrete, then

$$p_x(\alpha) = \sum_{i=1}^{n} P_{x|A_i}(\alpha|A_i) P(A_i). \tag{2.74}$$

Similarly, if x is a continuous RV then

$$f_x(\alpha) = \sum_{i=1}^{n} f_{x|A_i}(\alpha|A_i) P(A_i). \tag{2.75}$$

Proof. Let event $B = \{\zeta : x(\zeta) \le \alpha\}$, and define $B_i = B \cap A_i$. Then $\{B_i\}_{i=1}^{n}$ is a partition of B and

$$F_x(\alpha) = P(B) = \sum_{i=1}^{n} p(B_i) = \sum_{i=1}^{n} P(B|A_i) P(A_i)$$

from which the desired results follow.   ∎

Example 2.5.3.   *Resistors are obtained from one of two resistor manufacturers. Manufacturer 1 is event A_1 and manufacturer 2 is event A_2 with probabilities 1/4 and 3/4, respectively. Given the manufacturer, the conditional PDFs for the resistor values are known as*

$$f_{r|A_1}(\alpha|A_1) = 0.01(u(\alpha - 900) - u(\alpha - 1000))$$

and

$$f_{r|A_2}(\alpha|A_2) = 0.01(u(\alpha - 950) - u(\alpha - 1050)).$$

Find the PDF of the resistance value.

Solution.   From the Theorem of Total Probability, we have

$$f_r(\alpha) = f_{r|A_1}(\alpha|A_1) P(A_1) + f_{r|A_2}(\alpha|A_2) P(A_2);$$

hence,

$$f_r(\alpha) = \begin{cases} 1/400, & 900 < \alpha < 950 \\ 1/100, & 950 < \alpha < 1000 \\ 3/400, & 1000 < \alpha < 1050 \\ 0, & \text{otherwise.} \end{cases}$$

Drill Problem 2.5.1.   *A discrete RV x has PMF*

$$p_x(\alpha) = \begin{cases} \dfrac{1}{4}(0.8)^\alpha, & \alpha = 1, 2, \ldots \\ 0, & \text{otherwise.} \end{cases}$$

Event $A = \{\zeta : 2 < x(\zeta) < 5\}$ and event $B = \{\zeta : x(\zeta) \geq 3\}$. Find (a) $p_{x|A}(3|A)$, (b) $p_{x|B}(4|B)$.

Answers: 0.16, 0.5556.

Drill Problem 2.5.2. *The RV x has PDF $f_x(\alpha) = e^{-\alpha}u(\alpha)$, event $A = \{\zeta : x(\zeta) > 10\}$, and event $B = \{\zeta : -2 < x(\zeta) < 5\}$. Find $f_{x|A}$ and $f_{x|B}$.*

Answers: $e^{-(\alpha-10)}u(\alpha - 10)$, $e^{-\alpha}(u(\alpha) - u(\alpha - 5))/(1 - e^{-5})$.

2.6   SUMMARY

In this chapter, we have introduced the concept of a random variable. A random variable is a mapping which assigns a real number to each outcome in the sample space. Probabilities for events defined in terms of the random variable x may be computed from the CDF (cumulative distribution function) for x, defined by

$$F_x(\alpha) = P(\{\zeta \in S : x(\zeta) \leq \alpha\}). \tag{2.76}$$

For example, $P(a < x(\zeta) \leq b) = F_x(b) - F_x(a)$ if $b > a$, and $P(x(\zeta) = a) = F_x(a) - F_x(a^-)$. Any event of practical interest may be expressed in the form

$$A = \bigcup_{i=1}^{n} A_i, \tag{2.77}$$

where $A_i = \{x : a_i < x \leq b_i\}$ or $A_i = \{a_i\}$, with $A_i \cap A_j = \varnothing$ for $i \neq j$. Then

$$P(x \in A) = \sum_{i=1}^{n} P(x \epsilon A_i) = \int_A d F_x(\alpha) = \sum_{i=1}^{n} \int_{Ai} d F_x(\alpha). \tag{2.78}$$

Consequently, if the CDF is known, no integration is required.

If the CDF F_x is a jump function (piecewise constant), then x is a discrete RV, with PMF (probability mass function)

$$p_x(\alpha) = P(x(\zeta) = \alpha) = F_x(\alpha) - F_x(\alpha^-), \tag{2.79}$$

and

$$F_x(\alpha) = \sum_{\alpha' \in (-\infty, \alpha] \cap D_x} p_x(\alpha'), \tag{2.80}$$

where D_x is the set of points where $p_x(\alpha) \neq 0$.

If the CDF F_x contains no jumps then (for all practical purposes) x is a continuous RV with PDF (probability density function)

$$f_x(\alpha) = \frac{d F_x(\alpha)}{d\alpha} \tag{2.81}$$

and

$$F_x(\alpha) = \int_{-\infty}^{\alpha} f_x(\alpha') \, d\alpha' . \tag{2.82}$$

The RV x is a mixed RV if it is neither discrete nor continuous. The Lebesgue Decomposition Theorem can be applied in this case to separate the CDF into discrete and continuous parts.

The Riemann-Stieltjes integral was defined in order to provide a unified analytical framework for treating any type of RV. The Dirac delta function provides a useful alternative—enabling one to use a Riemann integral and a PDF containing Dirac delta functions in the mixed or discrete RV cases.

The conditional CDF $F_{x|A}(\alpha|A)$ was defined as

$$F_{x|A}(\alpha|A) = \frac{P(\{\zeta \in S : x(\zeta) \le \alpha \text{ and } \zeta \in A\})}{P(A)}, \tag{2.83}$$

along with corresponding conditional PDF $f_{x|A}$ and conditional PMF $p_{x|A}$.

2.7   PROBLEMS

1. Which of the following functions are legitimate CDF's? Why, or why not?

$$H_1(\alpha) = \begin{cases} 0, & \alpha < -1 \\ \alpha^2, & |\alpha| \le 1 \\ 1, & 1 < \alpha \end{cases}$$

$$H_2(\alpha) = \begin{cases} 0, & \alpha < 0 \\ \alpha^2/2, & 0 \le \alpha \le 1 \\ 1, & 1 < \alpha \end{cases}$$

$$H_3(\alpha) = \begin{cases} 0, & \alpha < 0 \\ \sin(\alpha), & 0 \le \alpha \le \pi/2 \\ 1, & \pi/2 < \alpha \end{cases}$$

$$H_4(\alpha) = \begin{cases} 0, & \alpha \le -4 \\ 1 - \exp(-a(\alpha + 4)), & -4 < \alpha \end{cases}$$

2. The sample space is $S = \{a_1, a_2, a_3, a_4\}$ with probabilities $P(\{a_1\}) = 0.15$, $P(\{a_2\}) = 0.25$, $P(\{a_3\}) = 0.4$ and $P(\{a_4\}) = 0.2$. A random variable x is defined by $x(a_1) = 2$,

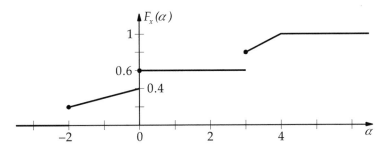

FIGURE 2.11: Cumulative distribution function for Problem 2.6

$x(a_2) = -1$, $x(a_3) = 3$, and $x(a_4) = 0$. Determine: (a) $p_x(\alpha)$, (b) $x^{-1}((-\infty, \alpha])$, (c) $x^{-1}([-1, 2])$, and (d) $F_x(\alpha)$.

3. Four fair coins are tossed. Let random variable x equal the number of heads tossed. Determine: (a) $p_x(\alpha)$, (b) $x^{-1}((-\infty, \alpha])$, and (c) $F_x(\alpha)$.

4. The sample space $S = \{a_1, a_2, a_3, a_4, a_5\}$ with probabilities $P(\{a_1\}) = 0.15$, $P(\{a_2\}) = 0.2$, $P(\{a_3\}) = 0.1$, $P(\{a_4\}) = 0.25$, and $P(\{a_5\}) = 0.3$. Random variable x is defined as $x(a_i) = 2i - 1$. Find: (a) $p_x(\alpha)$, (b) $x^{-1}((-\infty, \alpha])$, and (c) $F_x(\alpha)$.

5. Let

$$w(t) = \begin{cases} 2 - 0.5|t - 5|, & |t - 5| \le 4 \\ 0, & |t - 5| > 4 \end{cases}$$

Let $S = \{0, 1, \ldots, 19\}$ and $P(\zeta = i) = 0.05$ for $i \in S$. Let RV $x(\zeta) = w(\zeta T)$ for each $\zeta \in S$, where $T = 0.5$. Sketch the CDF for the RV x.

6. Random variable x has CDF shown in Fig. 2.11. Event $A = \{\zeta \in S : x(\zeta) > 0\}$, event $B = \{\zeta \in S : x(\zeta) \ge 0\}$, and event $C(\alpha) = \{\zeta \in S : x(\zeta) \le \alpha\}$. Find: (a) $P(x = -2)$, (b) $P(x = -1)$, (c) $P(0 \le x < 3)$, (d) $P(-1 < x \le 0)$, (e) $P(B|A)$, (f) $P(A|B)$. (g) Find and sketch $P(C(\alpha)|A)$ vs. α.

7. Consider a department in which all of its graduate students range in age from 22 to 28. Additionally, it is three times as likely a student's age is from 22 to 24 as from 25 to 28. Assume equal probabilities within each age group. Let random variable x equal the age of a graduate student in this department. Determine: (a) $p_x(\alpha)$, (b) $F_x(\alpha)$.

8. A softball team plays eight games in a season. Assume there are no ties, and that the team has an equal probability of winning or losing each game. Let random variable x equal twice the total number of wins in the season. Determine: (a) $p_x(\alpha)$, (b) $F_x(\alpha)$, (c) $P(4 \le x \le 12)$, (d) $P(2 < x \le 12)$, (e) $P(12 \le x \le 20)$, (f) $P(-1 \le x \le 12)$.

9. Which of the following functions are legitimate PDFs? Why? If not, could the function be a PDF if multiplied by an appropriate constant? Find the constant.

$$g_1(\alpha) = \begin{cases} 0.25, & -2 < \alpha < 0 \\ 0.5, & 1 < \alpha < 3 \\ 0, & \text{otherwise} \end{cases}$$

$$g_2(\alpha) = e^{-a|\alpha|}, \qquad -\infty < \alpha < \infty$$

$$g_3(\alpha) = \begin{cases} |\alpha|, & |\alpha| < 1 \\ 0, & \text{otherwise} \end{cases}$$

$$g_4(\alpha) = \frac{\sin(\pi\alpha)}{\pi\alpha}, \qquad -\infty < \alpha < \infty$$

10. Which of the following functions are legitimate PDFs? Why, or why not?

$$g_1(\alpha) = \begin{cases} 0.75\alpha(2 - \alpha), & 0 \le \alpha \le 2 \\ 0, & \text{otherwise} \end{cases}$$

$$g_2(\alpha) = \begin{cases} 0.5e^{-\alpha}, & 0 \le \alpha < \infty \\ 0, & \text{otherwise} \end{cases}$$

$$g_3(\alpha) = \begin{cases} 2\alpha - 1, & 0 \le \alpha \le 0.5(1 + \sqrt{5}) \\ 0, & \text{otherwise} \end{cases}$$

$$g_4(\alpha) = \begin{cases} 0.5(\alpha + 1), & -1 \le \alpha \le 1 \\ 0, & \text{otherwise} \end{cases}$$

11. For the following PDFs, find β, find and sketch the CDF, and then find $P(1 \le x < 2)$:
(a) $f_x(\alpha) = \beta\alpha^2 e^{-3\alpha}u(\alpha)$, (b) $f_x(\alpha) = \beta/(1 + \alpha^2)$, (c)

$$f_x(\alpha) = \begin{cases} \beta \sin(\alpha), & 0 \le \alpha \le \pi/2 \\ 0, & \text{otherwise} \end{cases}$$

12. Can a function be both a PDF and CDF? Why or why not?

13. The time (in years) before failure, t, for a certain television set is a random variable, with

$$f_t(t_0) = \frac{1}{5}e^{-t_0/5}u(t_0).$$

Determine: (a) $F_t(t_0)$; (b) the probability that the TV set will fail during the first year; (c) the probability that the TV fails after the 15th year; (d) $P(1 < t < 5)$.

14. The PDF for the time before failure for a piece of equipment is

$$f_t(t_0) = \beta t_0 \exp(-t_0/10)u(t_0).$$

Determine: (a) β; (b) $F_t(t_0)$; (c) $P(t < 10)$; (d) $P(2 \le t < 10)$.

15. Given CDF

$$F_x(\alpha) = \frac{1}{4}(\alpha + 1)(u(\alpha + 1) - u(\alpha - 2)) + \left(\frac{1}{2} + \frac{\alpha}{8}\right)$$
$$\times (u(\alpha - 2) - u(\alpha - 4)) + u(\alpha - 4),$$

Determine: (a) $f_x(\alpha)$, (b) $P(1/4 \le x < 3)$.

16. Given

$$f_x(\alpha) = \begin{cases} \beta\alpha^{1/2}, & 0 < \alpha < 1 \\ 0, & \text{otherwise.} \end{cases}$$

Determine: (a) β, (b) $F_x(\alpha)$, (c) $P(x \le 3/4)$.

17. Find the CDF for the following PDF

$$f_x(\alpha) = \begin{cases} (3\alpha - 1)^2, & 0 < \alpha < 1 \\ 0, & \text{otherwise.} \end{cases}$$

18. A fair coin is tossed twice. The RV x is the number of heads. Find and sketch the PMF and CDF for x.

19. Evaluate

$$I = \int_{-\infty}^{\infty} \frac{(9\cos(t) + e^{-t^2})\delta(t - 2)}{5t^2 - \tan(t - 1)}\, dt.$$

20. A PDF is given by

$$f_x(\alpha) = \frac{1}{2}\delta(\alpha + 1.5) + \frac{1}{8}\delta(\alpha) + \frac{3}{8}\delta(\alpha - 2).$$

Determine $F_x(\alpha)$.

21. A PDF is given by

$$f_x(\alpha) = \frac{1}{5}\delta(\alpha + 1) + \frac{2}{5}\delta(\alpha) + \frac{3}{10}\delta(\alpha - 1) + \frac{1}{10}\delta(\alpha - 2).$$

Determine $F_x(\alpha)$.

22. A mixed random variable has a CDF given by

$$F_x(\alpha) = \begin{cases} 0, & \alpha < 0 \\ \alpha/4, & 0 \le \alpha < 1 \\ 1 - e^{-0.6931\alpha}, & 1 \le \alpha. \end{cases}$$

Determine $f_x(\alpha)$.

23. A mixed random variable has a PDF given by

$$f_x(\alpha) = \frac{1}{4}\delta(\alpha + 1) + \frac{3}{8}\delta(\alpha - 1) + \frac{1}{4}(u(\alpha + 1) - u(\alpha - 0.5)).$$

Determine: (a) $F_x(\alpha)$, (b) $P(-1 \le x \le 0)$, (c) $f_{x|x>0}(\alpha|x > 0)$.

24. The random variable x has PMF

$$p_x(\alpha) = \begin{cases} 2/13, & \alpha = -1 \\ 3/13, & \alpha = 1 \\ 4/13, & \alpha = 2 \\ 3/13, & \alpha = 3 \\ 1/13, & \alpha = 4. \end{cases}$$

Event $A = \{x > 2\}$. Find (a) $F_x(\alpha)$, (b) $p_{x|A}(\alpha|A)$.

25. The waveform $w(t)$ is uniformly sampled every 0.1s from 0 to 3s, where

$$w(t) = \begin{cases} 3t^2, & 0 \le t \le 1 \\ 3, & 1 < t \le 2 \\ 9 - 3t, & 2 < t \le 3. \end{cases}$$

Event $A = \{w(t) < 3/2\}$ and event $B = \{0 < t < 1\}$. Let the random variable x be the sample value rounded to the nearest integer. Determine: (a) $p_x(\alpha)$, (b) $F_x(\alpha)$, (c) $p_{x|A}(\alpha|A)$, (d) $F_{x|A}(\alpha|A)$, (e) $p_{x|B}(\alpha|B)$, (f) $F_{x|B}(\alpha|B)$, (g) $p_{x|A\cap B^c}(\alpha|A \cap B^c)$, (h) $F_{x|A\cap B^c}(\alpha|A \cap B^c)$.

26. Suppose the following information is known about the RV x. The range of x is a subset of integers and event $A = \{x \text{ is even}\}$. Additionally, $F_x(0^-) = 0$, $F_x(1^-) = 1/8$, $F_x(4^-) = 7/8$, $F_x(4) = 1$, $p_{x|A}(2|A) = 1/2$, and $p_{x|A^c}(3|A^c) = 3/4$. Determine: (a) $p_x(\alpha)$, (b) $F_x(\alpha)$.

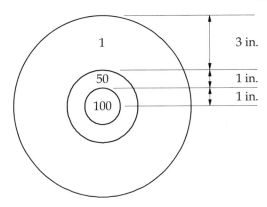

FIGURE 2.12: Target for Problem 2.31

27. Random variable y has the PMF

$$p_y(\alpha) = \begin{cases} 1/8, & \alpha = 0 \\ 3/16, & \alpha = 1 \\ 1/4, & \alpha = 2 \\ 5/16, & \alpha = 3 \\ 1/8, & \alpha = 4. \end{cases}$$

Random variable $w = (y - 2)^2$ and event $A = \{y \geq 2\}$. Determine: (a) $p_{y|A}(\alpha|A)$, (b) $p_w(\alpha)$.

28. Suppose x is a random variable with

$$p_x(\alpha) = \begin{cases} \beta\gamma^\alpha, & \alpha = 0, 1, 2, \ldots \\ 0, & \text{otherwise.} \end{cases}$$

where β and γ are constants, and $0 < \gamma < 1$. As a function of γ, determine: (a) β, (b) $F_x(\alpha)$, (c) $F_{x|x \leq x_0}(x_0/2|x \leq x_0)$.

29. The time before failure, t, for a certain television set is a random variable with

$$f_t(t_o) = \frac{1}{5}e^{-t_o/5}u(t_o).$$

Event $A = \{t > 5\}$ and $B = \{3 < t < 7\}$. Determine: (a) $F_{t|A}(t_o|A)$, (b) $f_{t|B}(t_o|B)$, (c) $P(B)$, (d) $P(A|B)$, (e) $f_{t|A^c \cap B^c}(t_o|A^c \cap B^c)$, (f) $P(A^c \cap B)$.

30. A random variable x has CDF

$$F_x(\alpha) = \left(\alpha + \frac{1}{2}\right)u\left(\alpha + \frac{1}{2}\right) - \alpha u(\alpha) + \frac{1}{4}\alpha u(\alpha - 1) + \left(\frac{1}{2} - \frac{1}{4}\alpha\right)u(\alpha - 2),$$

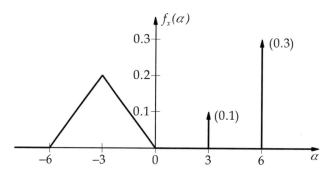

FIGURE 2.13: Probability density function for Problem 2.37

and event $A = \{x \geq 1\}$. Find: (a) $f_x(\alpha)$, (b) $P(0.5 < x \leq 1.5)$, (c) $F_{x|A}(\alpha|A)$, (d) $f_{x|A}(\alpha|A)$.

31. Judge Rawson, the hanging judge, does not treat criminals lightly. However, she does offer a pretrial sentence (in years) based on the outcome of a dart thrown at the target illustrated in Fig. 2.12.

    What the defendants do not know is that Judge Rawson is an incredibly accurate dart thrower. The probability of x years of sentence is the ratio of the area of the band marked $(100 - x)$ to the total target area.

    Determine: (a) the PMF for the sentence length from a dart throw.

    Three defendants choose dart sentences. Determine the probability that: (b) none of the defendants serve time; (c) exactly two of the defendants serve time; (d) each defendant is given a unique sentence.

32. The head football coach at the renowned Fargo Polytechnic Institute is in serious trouble. His job security is directly related to the number of football games the team wins each year. The team has lost its first three games in the eight game schedule. The coach knows that if the team loses five games, he will be fired immediately. The alumni

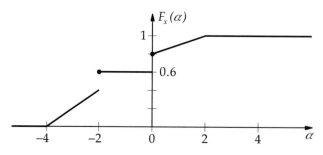

FIGURE 2.14: Cumulative distribution function for Problem 2.38

hate losing and consider a tie as bad as a loss. Let x be a random variable whose value equals the number of games the present head coach wins. Assume the probability of winning any game is 0.6 and independent of the results of other games. Determine: (a) $p_x(\cdot)$, (b) $F_x(\cdot)$, (c) $p_{x|x>3}(\alpha|x>3)$.

33. Consider Problem 32. The team loves the head coach and does not want to lose him. The more desperate the situation becomes for the coach, the better the team plays. Assume the probability the team wins a game is dependent on the total number of losses as $P(W|L) = 0.2L$, where W is the event the team wins a game and L is the total number of losses for the team. Let A be the event the present head coach is fired before the last game of the season. Determine: (a) $p_x(\cdot)$, (b) $F_x(\cdot)$, (c) $p_{x|A}(\alpha|A)$.

34. A class contains five students of about equal ability. The probability a student obtains an A is $1/5$, a B is $2/5$, and a C is $2/5$. Let the random variable x denote the number of students who earn an A in the class. Determine the PMF for the RV x.

35. Professor Rensselaer has been known to make an occassional blunder during a lecture. The probability that any one student recognizes the blunder and brings it to the attention of the class is 0.13. Assume that the behavior of each student is independent of the behavior of other students. Determine the minimum number of students in the class to insure the probability that a blunder is corrected is at least 0.98.

36. Consider Problem 35. Suppose there are four students in the class. Determine the probability that (a) exactly two students recognize a blunder; (b) exactly one student recognizes each of three blunders; (c) the same student recognizes each of three blunders; (d) two students recognize the first blunder, one student recognizes the second blunder and no students recognize the third blunder.

37. Random variable x has PDF shown in Fig. 2.13. Event $A = \{x : -3 < x < 6\}$. Find: (a) F_x, (b) $P(-5 < x \le 3)$, (c) $F_{x|A}(\alpha|A)$, (d) $f_{x|A}(\alpha|A)$.

38. Random variable x has CDF shown in Fig. 2.14. Event $A = \{x : -2 \le x < 4\}$. Find: (a) f_x, (b) $P(-2 \le x < 1)$, (c) $F_{x|A}(\alpha|A)$, (d) $f_{x|A}(\alpha|A)$.

Bibliography

[1] M. Abramowitz and I. A. Stegun, editors. *Handbook of Mathematical Functions*. Dover, New York, 1964.

[2] E. Ackerman and L. C. Gatewood, *Mathematical Models in the Health Sciences: A Computer-Aided Approach*. University of Minnesota Press, Minneapolis, MN, 1979.

[3] E. Allman and J. Rhodes, *Mathematical Models in Biology*. Cambridge University Press, Cambridge, UK, 2004.

[4] C. W. Burrill. *Measure, Integration, and Probability*. McGraw-Hill, New York, 1972.

[5] K. L. Chung. *A Course in Probability*. Academic Press, New York, 1974.

[6] G. R. Cooper and C. D. McGillem. *Probabilistic Methods of Signal and System Analysis*. Holt, Rinehart and Winston, New York, second edition, 1986.

[7] Wilbur B. Davenport, Jr. and William L. Root. *An Introduction to the Theory of Random Signals and Noise*. McGraw-Hill, New York, 1958.

[8] J. L. Doob. *Stochastic Processes*. John Wiley and Sons, New York, 1953.

[9] A. W. Drake. *Fundamentals of Applied Probability Theory*. McGraw-Hill, New York, 1967.

[10] J. D. Enderle, S. M. Blanchard, and J. D. Bronzino. *Introduction to Biomedical Engineering*. Elsevier, Amsterdam, second edition, 2005, 1118 pp.

[11] William Feller. *An Introduction to Probability Theory and its Applications*. John Wiley and Sons, New York, third edition, 1968.

[12] B. V. Gnedenko and A. N. Kolmogorov. *Limit Distributions for Sums of Independent Random Variables*. Addison-Wesley, Reading, MA, 1968.

[13] R. M. Gray and L. D. Davisson. *RANDOM PROCESSES: A Mathematical Approach for Engineers*. Prentice-Hall, Englewood Cliffs, New Jersey, 1986.

[14] C. W. Helstrom. *Probability and Stochastic Processes for Engineers*. Macmillan, New York, second edition, 1991.

[15] R. C. Hoppensteadt and C. S. Peskin. *Mathematics in Medicine and the Life Sciences*. Springer-Verlag, New York, 1992.

[16] J. Keener and J. Sneyd. *Mathematical Physiology*. Springer, New York, 1998.

[17] P. S. Maybeck. *Stochastic Models, Estimation, and Control, volume 1*. Academic Press, New York, 1979.

[18] P. S. Maybeck. *Stochastic Models, Estimation, and Control, volume 2*. Academic Press, New York, 1982.

[19] J. L. Melsa and D. L. Cohn. *Decision and Estimation Theory*. McGraw-Hill, New York, 1978.

[20] K. S. Miller. *COMPLEX STOCHASTIC PROCESSES: An Introduction to Theory and Application*. Addison-Wesley, Reading, MA, 1974.

[21] L. Pachter and B. Sturmfels, editors. *Algebraic Statistics for Computational Biology*. Cambridge University Press, 2005.

[22] A. Papoulis. *Probability, Random Variables, and Stochastic Processes*. McGraw-Hill, New York, second edition, 1984.

[23] P. Z. Peebles Jr., *Probability, Random Variables, and Random Signal Principles*. McGraw-Hill, New York, second edition, 1987.

[24] Yu. A. Rozanov, *Stationary Random Processes*. Holden-Day, San Francisco, 1967.

[25] K. S. Shanmugan and A. M. Breipohl, *RANDOM SIGNALS: Detection, Estimation and Data Analysis*. John Wiley and Sons, New York, 1988.

[26] Henry Stark and John W. Woods. *Probability, Random Processes, and Estimation Theory for Engineers*. Prentice-Hall, Englewood Cliffs, NJ, 1986.

[27] G. van Belle, L. D. Fisher, P. J. Heagerty, and Thomas Lumley, *Biostatistics: A Methodology for the Health Sciences*. John Wiley and Sons, NJ, 1004.

[28] H. L. Van Trees. *Detection, Estimation, and Modulation Theory*. John Wiley and Sons, New York, 1968.

[29] L. A. Wainstein and V. D. Zubakov. *Extraction of Signals from Noise*. Dover, New York, 1962.

[30] E. Wong. *Stochastic Processes in Information and Dynamical Systems*. McGraw-Hill, New York, 1971.

[31] M. Yaglom. *An Introduction to the Theory of Stationary Random Functions*. Prentice-Hall, Englewood Cliffs, NJ, 1962.

Printed in the United States
by Baker & Taylor Publisher Services